CW00864446

When the red horse spoke

Beth Duff

AuthorHouse™ UK Ltd.
500 Avebury Boulevard
Central Milton Keynes, MK9 2BE
www.authorhouse.co.uk
Phone: 08001974150

First published by AuthorHouse 9/17/2009

ISBN: 978-1-4389-8654-8 (sc)

This book is printed on acid-free paper.

For Chelsea – my wonderful red horse.

Acknowledgements

There are many two-legged and four-legged creatures who have helped me on this amazing journey.

It was attending the Associate programme at the Secretan Centre in Canada in November 2000 that moved the idea of taking my horses to work from being a vague dream to a real possibility with action plan attached. Special thanks go to Susan Nind and Ron Szymanski who encouraged me to make this 'a big dream.'

I am grateful to Kenneth McKay, our web designer, Lynn Clarke, our graphic designer and Liz Marchant, our PR guru. They took the time to be our 'guinea pigs' at the very start of our venture in order to understand what we were trying to do.

Shona Still and Callum McDonald at Loanhead Equestrian Centre looked after Chelsea and Susie for several years until we could have them at home. They could not have had better care – Callum and Shona treated them as if they were their own horses. They were also very supportive when we told them about **the red horse speaks**. Their hard work in the background contributed so much to the success of the launch and we continue to enjoy taking clients to Loanhead for workshops.

Ann Romberg and Lynn Baskfield of Wisdom Horse Coaching in Minneapolis have been a constant source of encouragement ever since we met them at a conference in Nashville. More recently I have specially appreciated their help with my research. They have also become dear friends as well as colleagues in this work. Barbara Rector, Ann Alden and the members of the EFMHA Research Committee have offered encouragement and also helped me with my research methods and writing. Lisa Walters has been very generous in sharing the research she has done with Dr Ellen Gerkhe. Su Wahl of Healing Arts Wellness Centre (HAWC) provided her wonderful facility and introduced us to the naturopathic methods that have become central to the research. I'd also like to thank everyone who has attended sessions at HAWC or who have been part of our house groups at conferences. You have all given me so much support and encouragement as well as being so generous in sharing you own learning. Special thanks go to Nancy Peregrine who has taken so many amazing photographs and for being very generous in allowing me to use them in presentations.

Thank you to Robin Gates for introducing me to the work of Carolyn Resnick and helping me to improve my own horse skills and to Fiona Adamson, my coach, who has done so much to support me on my inner journey.

None of this work would be possible without clients and I am grateful to all of them – especially those who have been with us since the early days. Thank you for having the courage to sell this idea within your organisation and trusting us and the horses to give you the learning you wanted. I hope we are now making it easier for you through the quality of our work.

At home, I am grateful for the support of the red horse community – Sue Stephenson,

Annette Brooks-Rooney, Andrea McColl, Ruth Vaughan-Hendry, Joy Wootten, Mercedes Jiminez and Magali Martinez, Emma Spence, Avril Oliver and Lucy Wilkins. Thank you for sharing the journey with me. Very special thanks go to Sue Hendry my regular horse person for being open-minded enough to give this whole project a go in the first place and to being such a great partner when we work together. I am grateful to Lisa Esslemont for looking after my diary, taking care of workshop logistics – and generally making my life much more ordered.

At home too, I am profoundly grateful to Aidan, my partner in business and life for his love and understanding both for me and our equine friends. I don't suppose for a moment he expected buying one horse would lead to this! It has been a joy to see him grow to love horses as much as I do and to know that when I am on my travels, he looks after them with great care and devotion.

Of course, none of this would have been possible without all the wonderful horses I have had the pleasure of meeting all over the world. They have all had so much to offer their human students and have taught with grace and just enough firmness to get their message across. Thank you Susie, Darcy, Storm, Bluebell, Katie, Milo, Beau, Sadie, Skipper, Hooper, Moon, Gypsy, Dove, Blaze, Fly, Digger, Tosca and all the others I have met on my travels. A very special thanks, of course, goes to Chelsea, my dear red horse. She is really the founder of **the**

red horse speaks. Not only did she fulfil my childhood dream, she inspired a whole new dream. Often when I tuck her up at night, I ask her whether she knew what she was doing when she started to teach me. Her soft, wise eyes tell me she always knew this is what we are meant to be doing – together.

Contents

Part I: *Inspiration*

Part II: *Perspiration*

Part III: *Science*

Part IV: Magic

Part V: In Conclusion

Part I: *Inspiration*

Chapter 1: The Meeting

The mare looked sad and old, but her face was kind. She looked as if she might once have been a good athlete. She stood quietly as I took off her boots, seemingly reassured by my touch, but not certain enough to respond with affection. Perhaps that was not surprising. I was later to discover that she had had four or five homes in the previous year and her athletic ability had been exploited, something that would cause her to suffer from arthritis at an early age. She was no pushover, though. The ragged clip marks on her neck suggested that she didn't like being clipped and had probably made that very clear.

Her kind face won my heart and in that moment began a very special friendship which has changed my life and that of my clients. She is Chelsea, who helped me to fulfil a childhood dream to own a horse.

When I was at primary school, we were always allowed a day off to go to the Royal Highland Show, held near our home in Edinburgh. My parents were of farming stock and considered a visit to the show to be 'educational'. Naturally, I agreed despite seeing plenty of fine livestock on my grandparents' farms.

I thoroughly enjoyed watching the judging, walking round various exhibits and even meeting people we knew - as long as they didn't talk for too long! But by far the best was watching show jumping in the Main Ring towards the end of the day.

I don't know what it was that made it so fascinating but I just knew I wanted to do it. My parents, however, had other ideas and so the piano lessons continued. Not that I minded too much but I still thought about the show jumping and even managed to practice without a horse. I used to tie a skipping rope between two trees on the back lawn as my jump, then I'd do my best canter up to the fence, collect up my horse/self a few strides out, launch forth, clear the jump and then canter on. I used to do this out on walks too, cantering up to imaginary objects and then jumping. Cracks on the pavement became water jumps and any small branch or log became an upright or spread. Sometimes there were even combinations.

None of this did my high jumping any good. Being tall and leggy, I was always good at that but horses go straight at fences and humans, in these pre-Fosbury flop days, came in from the side. It did, however, prove to be amazingly good practice for show jumping!

I learned to ride by reading most of the Pullein Thomson books. I knew (in theory!) exactly how to walk, trot, canter, turn, and stop. I also began to learn the basics about tack and grooming. Very occasionally, I got the chance to ride a real pony when we took visiting friends and cousins to Edinburgh Zoo where the Children's Section offered pony rides. Sometimes, too, I was lifted on to one of my grandfather's Clydesdales, by then retired from work on the farm, and allowed to ride bareback.

Eventually, I grew out of these summers of playing 'let's pretend' but I continued to enjoy watching show jumping, both live and on television, whenever I could and continued to enjoy 'horsey' books. School and music filled most of my adolescent days and at university I continued to mix academic effort with music and athletics. High jumping was no longer a strong point, though. I always found it hard not to canter up to the jump and that approach really does limit one's effectiveness - and takes A LOT of explaining to one's team mates!

After I married Aidan, we spent some time abroad and for a while lived at the foot of the Drakensberg Mountains in South Africa. Pony trekking was a popular pastime as riding up the steep slopes seemed to make much more sense than walking. Most of the resort hotels had ponies and so I began to ride whenever I could. The tough little Basuto ponies had definitely never read the Pullein Thomson books and generally ignored my attempts to determine their pace, but they were wonderfully sure-footed over rough and at times precipitous terrain. I did manage to practice trotting when they felt the time was right, but it was a bit bumpy! They were rather mean creatures, not above nipping occasionally and not nearly as friendly as the shaggy Highlands at home. Still, they were a start to my equine education as well as transport into some of the most spectacular scenery in the world.

Coming back to live in the South of England, buying a house and establishing my career put riding on the back burner again, except for occasional opportunities to go trekking which I enjoyed simply for the companionship of the pony and the chance to enjoy wonderful scenery.

In 1994, however, we fulfilled our ambition to move back home to Scotland. We built our house and the garden needed some effort as our plot was essentially a woodland clearing. It was a couple of years before we had much real free time, all the more so because I was working away from home as I continued a project I had started before our move.

Late in July 1996, however, I decided to forget work for an afternoon and go and ride at Balmoral. I'd been meaning to do this all summer and trekking there was due to finish at the end of that week. It was a case of go now or wait another year! Once Her Majesty arrived in August, the ponies took up their real duties – bringing stags down from the hills for the deer stalkers. I decided I just had to do it especially as it was a wonderful sunny afternoon.

I was allocated a lovely gentle Highland mare who gave me a super afternoon. We walked and trotted in reasonable harmony through woods, by the river and finally back in front of the castle. I had a smile that met round the back of my head and some very sore muscles! But it had been wonderful. I was hooked. When could I do this again? How could I improve my riding so I wasn't quite so sore? Her Majesty and her lovely pony have much to answer for.

I rushed back home, made a cup of tea and reached for the Yellow Pages. Was there anywhere nearby where I could ride? Would they think I was stupid wanting to learn to ride at the age of 45? Yes, there were stables just a few miles away, also in lovely countryside. Still feeling a bit foolish, I rang at once.

"Do you teach adults to ride?" I asked rather nervously.

"Oh yes. As a matter of fact, we have quite a number of adult beginners." The reply I'd hoped for.

I felt less stupid now. I explained that I'd done a bit of pony trekking but had never had any lessons.

"I suggest you book a couple of lunge lessons and see how you go from there. Come up and have a look around too, if you would like."

The visit was booked for the following day. The yard was just right. It was quite small but with good facilities, friendly faces and, most importantly, friendly, well kept horses. The lessons were booked to take place at once. Suddenly I was a six year old again! I was so excited.

In the Newsagents the following day, a child's pony magazine caught my eye. The cover advertised an article about lunge lessons. It was just what I needed. Never mind that it was meant for small girls. I was one again anyway!

My first lunge lesson was great fun. I was introduced to a beautiful large schoolmaster. I suddenly realised that I'd never actually ridden a horse since grandfather's Clydesdales, always ponies.

Having struggled not very elegantly up into the saddle, it was strange, but comforting, to have a substantial neck and head in front of me and plenty of mane to grab in emergencies! But it felt a long way up. We moved off and into our circle. It was just as described in the magazine, starting at walk with some exercises to loosen me up and help me to find my balance. We then moved up to trot. I discovered that although I knew the theory quite well, the practice was a little different and I was still being jogged along rather than truly trotting in time with the horse. But it got better quite quickly and soon I felt sufficiently relaxed to trot round with my hands on my head, on my shoulders and out to the side. By the end of the half hour, I had also managed to walk a bit without stirrups, in reasonable comfort, and trot rather less comfortably, gratefully hanging on to the mane!

All too soon our lesson was over, but I was allowed to walk my trusty steed back to his loose box or, more accurately, he ushered me back!

The next lesson followed a similar pattern but by now I was feeling much more confident and relaxed. The trotting was going fine. I simply listened to the horse's hoof beats and sat and rose to that. At the end of this session, I was deemed ready for my first hack, which was eagerly booked.

Come the day, I was a little alarmed to find that the other person hacking out was a very experienced, albeit rather rusty, rider. I hoped I wouldn't hold her back, but she was friendly and said she didn't mind at all. Our ride leader was also very kind and we set off into the forest. The gentle mare I'd been given couldn't have been kinder either and I soon relaxed and found myself chattering away to my companions in my usual fashion.

After walking and trotting for a bit, our ride leader suggested a short canter. She knew I'd not yet cantered but thought I would be secure enough to try a short burst. Hanging on to the pommel, I felt the horse push off into a wonderful rocking movement. I was bouncing about rather a lot, but it felt very exhilarating. After what felt like just a few strides, we came back to trot. I soon learned that this gentle mare always came back to trot when she felt her rider becoming unsteady! I got myself organised again and we set off for a few more strides. All too soon, the hour was up and we were back at the yard.

My next hack was promptly booked and thus a pattern became established. I always rode at the weekend and, when I was working at home, I rode out at least once during the week too. The hacks grew in length too from one to two hours as I got fitter and more confident. I soon passed the canter test and we kept going with the others. I also graduated to other horses, although my first gentle hacking school-mistress always remained a firm favourite.

The changing seasons are always a source of joy for me as each brings its own particular beauty. Riding out made me even more aware of this as we watched the trees take on their autumn colours, have their moment of glory and then fade into bare, often snow covered, branches until the spring. As a rider, I have also seen so much more wildlife, partly, I suppose, because humans are considered less of a threat when accompanied by a horse and also one is higher up and thus better able to spot things. I soon got used to seeing the resident heron, hearing the buzzards overhead and spotting deer.

As autumn gave way to winter, I also learned how to ride on ice – carefully and with no stirrups in case the horse slipped and I needed to get clear quickly – about the joys of wonderful silent canters through powdery snow, trying to avoid the snowballs tossed up from the hooves of the horse in front! By this time too, I'd acquired the basic gear – jodhpurs, hat, boots, chaps and gloves – and started to dip into horsy magazines and books to learn more.

I decided to do a week's trail riding plus the associated horse care by way of a spring holiday while Aidan golfed in Portugal. That was a trigger to increase my riding even more to get fit. Plans were laid for additional two hour hacks, a half day hack and a trial whole day just to see what it felt like and what I felt like during it and, of course, afterwards! This extra training was great fun, if somewhat chilly at times as we were riding quite high up during what was still, in our part of the world, winter. On our half day ride, we crossed high, open moorland with spectacular views but with a gale in our faces that came straight from the Arctic.

The whole day out was pleasant, but still cold. From the highest point of our ride, we could see the mountains to the north still clad with snow, and feel the icy wind blowing from even further north. We found a sheltered spot for lunch and I was glad I'd brought a flask of hot tea. I was tired after this one, but not stiff, and I quickly recovered. It made me realise that proper clothing, food and fluids are as important on day rides as they are on long walks.

Around this time, there was talk at the yard of looking for new horses for some people who wanted to buy. I remarked that I'd like a horse of my own within the next year or so. It seemed a long way off and there were others with a need more urgent than mine, so I didn't think much more about it. My casual remark, however, had not gone unnoticed.

Chelsea had come on trial for some-one else, but they had not got on well. One of the yard staff mentioned that she might suit me. I thought it would be good to try her out. I would need to do this anyway in order to have my own horse and so, at the very least, it would be good experience. It was Easter weekend and so I had plenty of time. I watched her being ridden by the ride leader as we hacked out on Good Friday. I liked what I saw and so I rode her out on a hack the following day. On Easter Sunday I had the thrill of going for a short ride all by myself. My riding ambition was to be able to ride out on my own with my own horse. Chelsea was very happy to go out on her own. By then I had fallen in love with her and decided that I would have her. Now all she had to do was impress Aidan. He had had no contact with horses – apart from the occasional meeting with a friend's pony as a child – but he does love animals. I was pleased to see him slip a packet of polo mints into his pocket as we left for this very important meeting. We walked over to the gate of her paddock and she came over to greet us. She nodded to me and then walked straight up to Aidan and nuzzled her nose into his neck. He too was hooked!

She wasn't an obvious choice for a first horse – not that well schooled and rather spooky – but even at that stage, there was a bond which went beyond any logic. I felt that although I had chosen her, she had also chosen me and so there was a partnership right from the start. She and Aidan seemed to get on well too. She was duly vetted, insured and paid for. I had a horse who was to change my life forever.

Chapter 2: *First Whisperings*

Now that I had my horse, understood at least the basics about caring for her and had support when I didn't know what to do, I thought that from now on I could just concentrate on my riding and enjoy her. I planned for her to be at working livery for this first summer. I was still away from home on a regular basis and this arrangement ensured that Chelsea got enough exercise and some schooling when I wasn't around. The people at the yard were keen to have her on this basis and I felt sure they would take care of her in my absence. It all sounded fine but I hadn't reckoned on Chelsea's thoughts on the matter! Basically she didn't like life in working livery and was perfectly prepared to make that clear. Anyone she didn't take to was either run away with or ejected. I got to the point of dreading my phone calls to the yard when I was working away from home. What had she done now? All the people who rode her were experienced and all those I met seemed very kind and patient but they divided into two categories. There were those who adored her and asked if they could ride her again (not many) and those who never, ever, wanted to see her again (the vast majority). I was puzzled but deeply touched by this. I was not very experienced and yet she never took advantage of me. We just enjoyed ourselves out and about. Clearly skill and horsemanship weren't the keys to Chelsea's heart, so what was?

Already she was whispering to me. She didn't like the present arrangements. I was away far more than I ought to be. I had to find a way of being at home more of the time and she was a real incentive for this. I loved the times I was at home that summer. I was up at 5.00am and riding by 6.00am. We met deer and their babies, watched buzzards hunt for breakfast but most of all just enjoyed one another. I had always been a bit of a workaholic. I loved my job, loved my customers and was proud of my business but Chelsea was whispering to me that there was more to life. I'd never had a really absorbing passion before. Yes, I enjoyed gardening, cooking and reading but all these could be laid aside for work. Chelsea was a much more powerful distraction. She had to be looked after and exercised and more than that, she had become a friend whom I missed if I didn't see her every day. She made me stop and think.

Something else made me stop and think too. Having a horse for the first time is a bit like having a new baby. Everyone wants to give you advice – and mostly it's contradictory. I discovered that, in the case of horses, there are three main schools of thought. There were the traditionalists, derived from the army and hunting, who thought that horses should be dominated otherwise they would take advantage of their riders. There was the racing world which was often harsh but tended to make some allowance for individuals, especially those who were winners. Finally there were the horse-centred people such as Monty Roberts.

Here the emphasis was seeing things from the horse's point of view and trying to understand their language.

Chelsea was not the easiest horse. Having had four or five homes in the year before I bought her, she had little reason to trust people. She also had a mind of her own and had learned to take care of herself – which was no surprise given her unsettled past. Out hacking, she would shy and spook and run away. I was always being told to "Take no nonsense" or "Make her get on with it". In my working life I'm a consultant and so I tend to use my influence as I generally don't have any direct authority over the people I work with. Just as I don't storm round my clients' offices telling them what to do, so I didn't feel inclined to take that approach with Chelsea. The traditional sirens were all about me, telling me she would soon be taking advantage of me but I decided to stick to my instincts to be kind but firm. I remembered the famous Birtwick Balls served up daily by John Manly, the kindly coachman in Anna Sewell's 'Black Beauty' – "kindness and patience, firmness and petting all wrapped up in common sense and fed daily."

It felt better to ask Chelsea to do something and praise her when she obliged, to recognise and understand her fears and to take time to earn her trust rather than whip her for being spooky. This approach was soon vindicated. Chelsea responded best when I was calm and confident. That way she was happy to let me be the leader. If I was hesitant and seemed bothered, she became alarmed and took charge. I'm sure that had I ever become angry with her when she was afraid, it would have terrified her even more. I tend to be a fairly calm person anyway so I soon found that I could cope by just being quiet if she was spooked by something. The effect this had on her was remarkable. At first, if any small thing alarmed her early in the day, she spent the rest of the day expecting monsters round every tree and corner. Gradually she became much calmer. These days, scary things cause less alarm and are quickly put behind her – and she's stopped walking round the outside bend on tracks to see what is coming! I've noticed too that the whites of her eyes, so often visible in these early days, are now rarely seen and her eyes have become soft and gentle. A quiet word of encouragement and a pat for being brave works wonders. Undoubtedly, under Chelsea's tutelage, I was becoming even more unflappable.

I was fascinated by this learning from Chelsea and thought that it would be great to share it with my clients. At the time, I was involved in a huge change management project in Slough and was coaching the rising stars who would be amongst the senior managers in the new organisation. I could see how this newfound wisdom could help them develop their leadership skills but I didn't, at that stage, have a means to get my message across. However, the idea of **the red horse speaks** was born. I continued to be open to Chelsea's learning and to play with the idea of creating a programme with horses whilst I thought about how I could make space in my life for this new development.

I was deeply involved with my project and was very fond of all the people with whom I was working and so the work with horses would have to wait until this was completed.

In the meantime, I had plenty more opportunities to practice my own skills out and about with Chelsea, especially on the days when her brakes were non-existent. It wasn't that she bolted in panic, more that she took off and wouldn't stop. This was my first attempt at following the 'Monty Robert's approach'. I was much inspired by Monty's book 'The Man Who Listens to Horses' which had been published not long before I got Chelsea. In the book, he describes his experience with a horse called My Blue Heaven who also would not stop. Monty taught the horse to love stopping more than keeping going and before long they were winning competitions. My riding skills are no match for Monty's but I figured that I could try to follow the principle. We started at walk. When I asked her to halt and she responded, she got a pat and lots of praise. Gradually we worked up to trot and then, greatest test of all, canter. It was a while before I could really trust the brakes but we got there in the end and these days she stops simply in response to a quiet voice request.

Chapter 3: *Getting to know each other*

A couple of months after Chelsea came into my life I joined a group of ladies for a three day expedition through the hills. Chelsea had never done anything like this before but I hoped she would enjoy it. I was still fresh after my week of trail riding and hoped Chelsea and I would be able to do long rides together. She took to it with what I was soon to know as her typical joie de vivre and enthusiasm for anything that entailed being out and about.

On the first day, she cantered up a mountain – her choice of pace – and confidently crossed rough, plank bridges. Neither of us has a head for heights so we soon struck a deal. She would carry me up hills and I would get off and lead her down. So good was she to lead that I was able to negotiate a burn by leaping between stepping stones and allowing her a long rein confident that she would stay with me and not pull me into the water. Her willingness to stay with me was essential during our second day out. Our path had been washed away by spring storms and we were forced to follow a narrow deer track and then find our own way down the steep hillside back onto our track below. We found a patch of blaeberries and struck off straight down what seemed like a wall of death. I kicked my heels in to cut out steps. Chelsea looked at me as if to say "It's OK, I'm right with you". It was a look that I would get to know well when things were tricky. All the way down the hill, she slipped and slithered in perfect time with me. Had she hesitated or gone ahead, she would have pulled me off my feet. She was a star. She got a big hug at the bottom and a polo mint. I had a grateful sip from our communal hip flask. It had been a difficult moment and I was touched that she had trusted me so much.

It was on this expedition too that I discovered her cleverness in assessing the going. We came across a large, very boggy area. It was particularly bad because of all the spring storms but it must have been bad anyway as a very narrow, marked track had been provided for walkers. It was not suitable for the horses, though. Our leader struggled across the bog on her pony but, having watched this, the rest of us decided it was better to let the horses go one by one to pick their way across, unhampered by our extra weight, while we used the track. The ponies plunged their way over, up to their knees in places as they ploughed across. It came to Chelsea's turn and someone remarked "Poor Chelsea! She won't have a clue what to do." 'Poor Chelsea' raised her beautiful Thoroughbred head and imperiously surveyed the scene, then her Irish Draught instinct took over and she picked a delicate path through the bog, well wide of the direct route and emerged to meet me with barely muddy hooves.

I later came to recognise this as her ancient wisdom, learned from the great collective unconscious of Irish Hunters who are so renowned for their ability to carry their riders safely over any ground. It got me wondering about my own ancient wisdom. I noticed that I always seemed to know how to deal with all the minor ups and downs that go with horse ownership. I come from a long line of farmers on both sides of my family and many had worked with horses. Although I had grown up with very little direct experience of these amazing animals, with Chelsea's help I now felt as if there was something deeper to call on... and this got me curious about the collective unconscious and other related topics, inspiring me to read the works of Rupert Sheldrake, Lynne McTaggart and others.

This sense of ancient wisdom was later to inspire my coaching too. I have had the privilege of working with individuals who were at critical points in their lives, asking the BIG questions. "What next?", "What else?", "What am I here to do?" Coaching with the horses has inspired all these clients to find their own ancient wisdom, to tune into the universe and to follow their passion.

I couldn't help but be aware of Chelsea's deep knowing about her surroundings. She missed nothing and I soon learned to be much more observant. One morning we were out very early, merrily trotting along, enjoying the spring sunshine. The land fell away to one side of our track and was rough and heathery. On the other side, it rose up and was thick with pine trees. Suddenly, from nowhere, a roe deer jumped out of the heather, ran under Chelsea's neck and darted into the trees. I got such a fright that my rising trot rose very high indeed and even from that great height I could see every hair on the deer's back! In the meantime my often spooky mare had simply checked her stride momentarily to let the deer through and continued on as if nothing had happened. She was obviously completely aware of the deer's presence and probably equally aware that I wasn't and so she simply took action to deal with the situation.

We continued to enjoy our summer together in between our respective 'work.' Inevitably, she got the occasional scratch and bite and my knowledge of equine first aid improved. From the start, she seemed to understand that I was helping her, even if it was a bit uncomfortable. She never ever tried to bite or kick, although she would occasionally flinch. On one occasion, she had a slightly runny eye and I was given a tube of cream to put in it. I was told to tie her up and then get someone to hold her head whilst I squeezed some cream into her eye. I actually found that I didn't need any help. She just let me put the tube into the corner of her eye and apply the cream underneath the eyelid. Normally Chelsea was appalled at any interference with her head and would jerk backwards in alarm. Some special wisdom seemed to tell her that this intervention was in her best interest. On another occasion, she developed an abscess on a front foot. After poulticing it to remove any deep infection, I washed out her hoof twice a day and applied Stockholm tar to keep it clean. She soon got to know the routine and could be relied upon to stand with her hoof on a towel to keep it clean whilst it dried out, even when

I left her unattended to go and wash out the buckets I had used.

I was surprised at the strength of the bond that developed between us. At first, I was usually away for a week at a time, but she always seemed to know she was mine. I would aim to arrive at the yard as she was finishing her final working hack. When she came back in, she would always walk up to me and stop. I was deeply touched by this as I felt it was more than I deserved.

I also began to realise just how sensitive she was to how I was feeling. One weekend I came home with a very heavy cold. I was weak but felt that a bit of gentle fresh air would be good for me. I thought I might manage a quiet hour out with her as I didn't feel I had the strength for anything more. From the moment I went to get her in from the field, she knew I was not myself. She came straight up to me and walked in with me very gently. Had she been a person I'm sure she'd have been taking my arm! I tacked her up and set off. That day she was my transport and carried me round. We stayed out for two hours and it was just the tonic I needed. We did no more than walk and trot but, when I tired, she came back to walk unasked. When we reached the tracks where we often cantered, she gathered herself ready to go, but I told her I was just too weak and so she continued to walk with good grace. She got a specially big hug when we got home.

I also came to understand the bonds that develop between horses and how little we consider this as horses are bought and sold.

I had just turned Chelsea out after a ride. It was winter and she had a cosy rug on. As ever, after turning her out I watched her for a bit. I always enjoy doing this. Some days she literally hightails off to her friends; other days, it's head down to eat. Today a roll was a high priority. She had her usual energetic roll, going over completely from one side to the other. Imagine my horror when she rose and I realised that the fastening on her rug had snapped and the whole rug had now slipped over her quarters. I ran to get a lead rope and walked up the field as calmly as I could, telling her that I was coming to help. As I walked towards her, her field mates decided it was time for a madcap gallop. My heart was in my mouth as I had visions of trailing rugs and broken legs. Amazingly, she didn't move. Another mare with whom she was friendly stayed with her, ignoring the high jinks all around. It was incredible to see these two, standing quietly in the middle of the riotous behaviour all around. I snapped on the lead rope, unfastened the rug and lead Chelsea in to find another rug. Only when I had led her away did her friend join in the fun. I was later to discover that this special friend had spent many years sharing her field with a succession of rescue horses and ponies whom her owner had adopted. She treated them all with the same kindness and sense of awareness that she had shown to Chelsea.

My contract in Slough was due to finish the following April and I resolved to ensure that future work could be undertaken either locally or, preferably, done from home as much as possible. I had also resolved to work less than full time, averaging about four days a week rather than five or even six. This was quite something for me. Hitherto I'd been a happy workaholic but now I was realising that there really was more to life and I was ready and willing to make some changes. Besides which, there was still that idea of bringing Chelsea into my working life.

Chapter 4: *New Beginnings*

The run up to finishing in Slough was inevitably hectic and I was away from home from Monday to Friday for several weeks. There was still time for some fun, though, as Chelsea and I made the most of weekends, even when the weather was bad. She never minds the weather, whatever it is. Most horses will try to protect their faces from heavy rain and snow, bending like bananas or even spinning round. That's not surprising as left to themselves they naturally put their backs to the wind. Chelsea, however, simply goes straight into any weather. We had some fun hacks in the snow. Good, fresh, powder snow is no particular hazard for horses and there is something magical about cantering through fresh snow on a winter morning in muffled silence. I also found that, just as she had assessed boggy ground so well in the early summer, so also was she excellent in snow and ice. I'd love to know how she always understood which icy puddles will break beneath her hooves and which will not. I soon came to trust the way she dealt with any icy stretches we found. She would gather herself in balance and place me carefully so that she could carry me safely. If she thought it was too slippery to be safe, she simply stopped and I knew it was wiser to get off and lead her. She also told me on the occasions that her hooves got balled up with snow. I soon became sensitive to this and dismounted to clean them out with a hoofpick when necessary.

During this winter, I also had my first chance to jump. When I'd bought Chelsea, I knew my own riding skills were very basic and I tried to have lessons regularly. Usually I had my lesson with Chelsea, but occasionally I borrowed a school horse as sometimes it was easier to learn this way. I still wanted to jump but I knew that Chelsea was just too powerful and athletic for me to start with. She cleared everything by miles and even some of the more advanced students at the yard felt they were having flying lessons rather than jumping lessons with her. The lovely school horses came into their own once again. My first real jump will stay etched on my mind forever. I was riding a beautiful, kind, grey mare who knew exactly what to do without any help from me. The little cross pole was probably no more that about eighteen inches high but when we turned in to it, I knew exactly what it was going to feel like. It was utterly uncanny. I had done this in my mind hundreds of times as a little girl and now it was happening for real. We soared over the jump and the smile on my face met round the back of my head. In that moment, I believed even more all that I had read about the power of the mind and the imagination. When it was too icy to ride out, having a jumping lesson became a great alternative and I gradually progressed to small combinations and slightly higher fences. My jumping is far from brilliant but I always know what to do, even if I can't always execute it quite as well as I can envision it. And, I always thoroughly enjoy the experience!

All this stood me in good stead for an opportunity that was to have a profound effect on my thinking – and indeed my approach to life. Aidan had been on a course with Jack Black, a fellow Scot who had a sound reputation for motivating individuals and businesses to achieve great things. He had worked with athletes and football players as well as large organisations. Aidan thought I would like it and he had the opportunity to take a repeat of the course for the princely sum of £10.00. We went off down to Edinburgh and had two wonderful days learning many exercises and techniques to improve performance in all aspects of life. It was perfectly timed for me as the project in Slough was coming to and end and the course inspired me to dig deep and finish well. It also gave me some time to think about the future I wanted afterwards – a future that definitely included Chelsea.

At last it was the end of April. I finished in Slough, leaving the people there in good heart and settling happily into their new organisation. I came home for a summer free of work. At least that was the plan. Some of my riding plans had to be altered a bit, though, as I had slipped a disc in my back at the end of February. As with so many injuries, it was a silly thing – I slipped on a piece of paper on the floor. It didn't really matter, though, as I had lots of ideas to try.

First among these, was to try 'Join Up' and loose schooling. I had read Monty Robert's book several times and had been fortunate enough to see a demonstration by Richard Maxwell who has a similar approach to Monty.

Now that I could be up at the yard early in the morning, I could use the outdoor school before it became busy with lessons. It was bigger than ideal, but the extra running round kept me fit and my back benefitted from the exercise too. It was, in any case, pure joy to be out of doors on these early summer mornings after being confined to an office for most of the winter. Chelsea got the idea of 'Join Up' very quickly and was soon walking round beside me. I was intrigued to find that if I jogged, she trotted and if I stood on the spot and turned away, she would come round with me. It soon became a great game which we both enjoyed enormously. I gradually taught her then to move away from me and to work loose in a circle. This required some precision on my part as I needed to look at her in just the right way and focus on the right spot on her quarters to keep her on a circle round me and not to go racing off to the other end of the school. Soon she was walking, trotting and cantering round me and working off my voice but our favourite part was always when I dropped my head. She knew then that she was free to come and stand by me while I stroked her forehead.

I was amazed at the strength of bond that this had created. One afternoon, I was playing with her in the partially covered arena which was near a small sawmill. I spotted a large lorry full of gravel come up the road to deliver its load. The clatter of the gravel being dumped was guaranteed to give Chelsea a fright. When the gravel hit the ground, she was standing near me and much to my amazement, though she was startled and lifted her head in alarm, she just stepped nearer to me with me between her and the noise. She then stood quite still. I spoke quietly to her, patted her and told her how brave she was.

I was amused that she should make sure I was between her and the noise, but I soon found out that this was quite deliberate on her part whenever there was something she was unsure about. Over the years, we have met bonfires, large farm machines, timber vehicles, other large or noisy objects and even pigs. She has never baulked at going past them, even if she was afraid, but she always insisted on having me between her and the scary object. She clearly trusts me in these moments and is very happy to accept my leadership.

In that summer, free of the pressures of work, I often reflected that Chelsea was herself an interesting experiment in applying leadership. If I was calm and positive, she would follow me anywhere. She certainly wouldn't tolerate any form of bullying or force. She occasionally needed time to just think about things and then make up her own mind as to whether what I was asking was, indeed, a good idea. When I wanted to teach her something, clear, consistent instruction got the message across quickly and easily. When she seemed not to understand, I tried to put myself in her shoes and consider the situation from her point of view. It reminded me that we all see the world with our own, different perspective. Just like the people in my project teams, she responds very well to kind words, praise and encouragement. I had always trusted my instinct but Chelsea helped to remind me that this was generally quite accurate. Most of all, I came to realise that gentleness goes a lot further than brute force or coercion and is really a great strength.

There has never been any doubt in my mind that she has made me a better leader. Having become used to watching her body language – a horse's main form of communication – has made me sharper at reading the unspoken messages that people often give. I have also found that people respond to 'Join Up' too. It is possible to 'send people away' and make it uncomfortable for them when their behaviour is not what you want. Equally, it's easy to make them feel comfortable around you when they are doing all the right things. I have to be careful not to appear manipulative in applying this but, with a bit of thought, it works well. I'm still practicing!

During that lovely summer, Chelsea had her first attempt at cross country. I had been riding her quietly due to my back problems but was happy to share her with a young lad whose own horse had to be retired due to age and injury. He was about to start university so it wasn't the moment for him to look for a permanent replacement. He loved Chelsea – and it was mutual – so he borrowed her for his last summer of pony club activities. They went well together on the flat and over jumps. Chelsea's love for jumping was always apparent and inspired great confidence in her rider.

I was intrigued to watch her first attempt at cross-country. Her young rider was very experienced and there was a little course of cross country jumps in a field near the yard. They popped over the small hedge then the big log a couple of times just to get warmed up. Chelsea was already beaming with pleasure. They came to the first combination fence – a small step down, a jump across a small ditch then a step up the other side. The first attempt was a bit ungainly with Chelsea just being firmly pushed through each element, but as they came round again, I could see her looking quite intently and planning her attack. This attempt was more fluent. The next time, however, she knew exactly what to do. The concentration was intense, although the smile was still there too. I could almost see her thinking "Down, over and out" and I certainly knew exactly where she was going to put her feet. It was perfect and they cantered on to the next fence which was a small water jump that gave her no problems at all.

The next fence, though, was more of a challenge. It was two upright rustic fences with just a bounce between them. Show jumping courses don't have fences that close together. There is always at least one full canter stride between combination fences on a show jumping course, so we didn't know what Chelsea might do. Our concern was that she might think it was a big spread fence and try to take both elements in one go. After all, spread fences are very familiar to show jumpers and Chelsea has great scope over them. They approached the fence at a steady pace so that Chelsea would be encouraged to think of it as two fences not one big one. She jumped the first element and hesitated as she realised there was no room for a stride. She was moving forward with such impulsion that she nudged the second element with her chest. Undaunted, however, she took a huge cat leap and cleared it somehow from a standing start. Horse and rider landed on the other side in one piece – just!!

What happened next was another remarkable insight into Chelsea's character. There were other horses grazing in the field. They had wandered up to watch Chelsea but weren't in the way. Chelsea wasn't bothered by their presence as she was having too much fun jumping. However, when Chelsea made her rather ungainly leap out of the bounce fence, the others seem to feel her unease and raced off down the field. Chelsea herself was a bit startled by the experience and her rider had absolutely no control over her, having lost reins, stirrups and balance. No-one could have blamed her if her herd instinct had taken over and she'd raced down the field with the others. But she didn't. When she landed, she stopped and stood quite still while her rider got himself back together again. They finished off their session going back over the little warm up fences to give them both a confident finish. The two of them continued to have a lot of fun with cross country fences that summer and ended up winning the local hunter pace competition in the autumn.

Chelsea was so much happier that summer. She is definitely a one-person horse. All the staff at the yard had remarked to me how much nicer and easier she had been since I had been able to spend much more time with her. I decided I was probably a nicer person too when I spent time with Chelsea and didn't have to travel so much. I had planned to have a four month break, hopefully beginning to pick up work again from September. I still loved what I did and had plenty of appetite for more but it was no longer the 'be all and end all' it once had been. I did a lot of thinking and decided to look for assignments that I could do mainly from home but September was still some way off – so the thoughts were occasional and still forming.

My summer of rest, however, came to a somewhat sudden end with a phone call from someone I had worked with some years previously. He wanted to propose me for a job in London. Would I be interested? To be honest, I wasn't really. Apart from being determined to enjoy my break, we were also selling my parents-in-law's house. They had moved into a nursing home some time before and had now decided that they wanted to sell their old home. There was a lot of clearing out to be done and we wanted to take our time, thinking what to sell and what to keep. I agreed to be put forward for the job as long as I didn't have to start until after the house had been handed over to its new owners.

The project, which was scheduled to last for just six weeks, was to help a consortium of health and local authorities in a bid for Health Action Zone status. I decided to do the work and then continue with my break. I had now become quite well practiced in the positive thinking techniques I had learned on the Jack Black course and decided to be bold and ask for what I wanted by way of terms and conditions. Previously I had a tendency to undervalue myself and be rather too accommodating to the needs of others. I didn't want the earth but it was now quite important to me to earn enough and still have time for Chelsea. The rate I requested was accepted and most of the work could be done from home. As a bonus, the person I worked with most closely had owned ponies and horses when she was young and enjoyed hearing about Chelsea. She trusted me to put in the effort required and didn't mind the occasions when I was out riding when she phoned.

I enjoyed this new way of working. I had always done quite a lot of travelling but the pattern I had developed when working in Slough was two weeks away and two weeks at home, although I was always home at weekends. I had started this routine when we were looking after Aidan's parents while they were still living in their own home. It meant that I could give Aidan a break when I was at home. Although we enjoyed looking after them, it was quite hectic on top of a busy work schedule as we had to cook for them several times a week, do their laundry and, on a daily basis, manage their medication. Although they eventually became so frail that they had to be admitted to a nursing home I continued with my routine till the last few months of the project when I eventually had to be in Slough most of the time. Now, however, I was at home most of the time. Most weeks I went down to London for the day although sometimes I stayed over when things were busy. I soon got to know all the others involved in the project and really enjoyed working with them.

To everyone's delight our bid was successful and I was particularly pleased that my part of the bid had been awarded all the funding I had requested. I hadn't really expected that as it so rarely happens. What I expected even less was to be invited to stay on and manage the implementation of the project. I asked whether my being based at the other end of the country might be a problem but everyone had decided that I was just as accessible as everyone else through phone, fax and e-mail. Moreover they felt that my not belonging to any of the partner organisations was a positive advantage. That was the end of my holiday but I had managed to sort out a routine which was almost ideal. It was a project to which I felt very committed, was not quite full time and could be done mainly from home.

It was while I was I working in London that I really became aware of how much Chelsea was teaching me. On the days I went down there, I spent most of my time in meetings but if I needed a space to work, I 'hotdesked' in my project sponsor's office. I didn't know the people there very well as they were involved only at the fringes of the project although we exchanged the usual pleasantries in the kitchen and at the photocopier. Gradually, however, they remarked that they enjoyed the time I spent in their office as they were sure I brought an air of calm with me. I was pleased to help as I thought they always seemed so frantic and overwrought. At one point, they knew that a particularly difficult week was approaching and they asked me if I could work in their office just so the place would 'feel better.' I would have preferred to be at home but agreed to help them as I had got to know them well by this time and was fond of all of them. I have no doubt that it was Chelsea's teaching me to be aware of my energy, to be calm and consistent that enabled me to help in this way.

Chapter 5: A Change for Chelsea

Through the autumn and winter, we settled into our new routine. Occasionally if I had to be away for a few days, I arranged for one of the yard staff whom Chelsea liked to ride her. Sometimes she was the lead horse on a hack, sometimes she was schooled. She didn't mind this as she knew who was riding her and so was much more settled.

Chelsea has an uncanny way of judging people. She was always very clear about the riders she liked and those she didn't, with the latter very often ending up on the ground! She had the same knack with people on the ground too. Some she totally ignored or avoided. Others she would allow to admire her and stroke her lovely velvet nose. She was particularly fond of children and could be left safely in the charge of the smallest child whilst I made any last minute preparations for a ride. Sometimes, as I kept a watchful eye, I was never quite sure who was looking after whom.

One afternoon, I had brought her in and was busy giving her a good brush when a family appeared outside the stable. I went out to ask if they needed help as I thought perhaps that they had missed the road back to the car park. I found parents and grandparents with two very disabled children, a little girl of about five or six in a small wheelchair and her slightly older brother who was able to walk.

"Are there any horses in?" enquired the father.

"Only mine" I replied, rather dreading his next question.

"Could we come in to see it?" I knew he would ask.

"Yes, of course" I replied, not wanting to disappoint them but expecting that when Chelsea heard the little girl, who was by now squealing with delight, and saw the wheelchair, she would retreat to the back of her box.

I thought to myself that as the doors were tubular metal, the little girl would at least be able to see her through the bars. Chelsea amazed me. As soon as the grandparents, mum and daughter arrived at her box, Chelsea unhesitatingly came over to the door and gently put her head down to greet the child. The little girl, who had no power of speech, continued to squeal with delight and wave her hands about. My lovely Chelsea, who would not have tolerated such behaviour from any normal person, clearly understood that this little human being was simply expressing her delight in the only way she could and continued to nuzzle her. I had some polo mints in my pocket and I knew Chelsea could be trusted to take them gently and so I put one into the little girl's hand. Her fingers were bent and so she couldn't offer Chelsea the mint on the open palm of her hand as one would normally do. Chelsea didn't mind. She simply nuzzled her way to the mint and took it very gently from between the tiny bent fingers.

By now, her brother had decided to be brave and come in. I asked him if he wanted to give Chelsea a polo mint too. He did and offered it to her on the palm of his hand. He wasn't completely sure of her and his hand was being tickled by her whiskers so he couldn't keep it still. This didn't worry Chelsea either. She just moved in time with him, back and forth, until she could gently take the mint. The little boy was laughing, no longer afraid now, just enjoying the tickling sensation of her whiskers.

There was a big lump in my throat. Chelsea really had made the afternoon for these children, especially the little girl who clearly loved horses. Chelsea had known that they meant no harm and that they needed her love and affection. She had seen the real people beneath the apparently disturbed behaviour. It was the first inkling of just what a sensitive teaching horse she would be.

During the winter, I had a further example of her strong attachment to me on an occasion where she might understandably have been more attached to another horse. I was bringing her in from the field one evening along with her best friend, the lovely chestnut mare who had looked after me so well when I began to ride. I often smiled to myself as I watched these two together in the field, never far from one another and often standing together by the hay ring looking for all the world like a couple of city girls enjoying a drink at the end of a busy day in the office.

It was winter and the road down from the field was icy. We picked our way to the stable block, me in the middle with a horse on either side. Something spooked the other horse – even now I have no idea what it was. It was unusual for her as she was normally so calm but suddenly she was dancing at the end of her lead rope on her back legs, trying to reverse at speed. I looked at Chelsea somewhat anxiously. If she followed her friend's lead I was not at all sure I would be able to keep my balance on the ice. I needn't have worried. It was 'that look' again and she was standing, head down and completely calm.

"Don't worry about me" was her clear message.

"Good girl," I whispered back to her, "Just stand quietly".

Her friend quietened enough to have four feet back on the ground but she was still alarmed. "Walk on" I said. Chelsea responded instantly and our combined forward momentum persuaded the other mare to come with us, albeit at a rather nervous jog. We got to the stable and I put Chelsea's rope through the ring outside without tying it. I knew she'd not go far. Usually I could trust them to walk in together but tonight I took no chances – especially as I still didn't know what had caused the fright. I led the other horse into her box, closed the door and turned to get Chelsea. I smiled when I saw her. She had moved over just far enough to be able to see into the stable. She must have wanted to watch and make sure all was well. I patted her and pulled the rope out from the ring and led her to her box, deeply touched that she had chosen to follow me rather than her friend.

During the worst of the snowy weather, I continued to have lessons. My jumping was coming on and I now felt ready to have a go with Chelsea. It was such a thrill. We simply trotted up to the jump, took one canter stride and popped over. She had more impulsion even in trot than most horses I'd experienced had in canter and there was no doubt that she really did jump for joy. It was an exciting moment and I loved to feel her energy and enthusiasm even although it sometimes led to her jumping me right out of the saddle and on to the floor! Happily the indoor school made for a soft landing and dear Chelsea always stopped dead in her tracks when I fell off. She planted all four feet where they were and wouldn't move till I was back on my feet.

Our jumping came to an end, however, all too soon. I had begun to notice as I rode her out that she did not always go forward with great enthusiasm and that occasionally she would jog rather than walk. This was not like her and I sought some advice. "Oh, just kick her and tell her to get on with it. She's just being lazy". I ignored that advice. I knew my equine friend too well by this time. If Chelsea was being lazy, it was for a reason. I continued to keep an eye on her for a week or so and she seemed no better – though no worse either. I asked one of the freelance instructors whom I liked and trusted to ride Chelsea and make an assessment. Chelsea had bucked whilst being asked to canter in the school and the instructor was of the opinion that it was "a stiff buck, not a naughty buck".

Despite being told by some of the staff that I was wasting my money, I asked the vet to come and examine her. Not for the first time, I was glad that I had listened to Chelsea and not the siren voices around me. The vet found that Chelsea had spavins in her hocks. This is an arthritic condition that in a horse as young as Chelsea, then only 9 years old, probably meant that she'd been over-jumped as a youngster. The vet also thought that she might have had a back injury at some stage, also consistent with jumping. It tallied with what we already knew of her in terms of her ability as a show jumper. What a pity some-one had seemed to take advantage of her. Perhaps this is why she had ended up in the hands of dealers. Had some-one not wanted to or not been able to nurse her back fully from injury? I had had her carefully checked when I bought her and although she had passed the vetting, there had been some doubt about her hocks but not enough to stop me buying her, especially as I didn't have any major sporting ambitions and had, in any case, already fallen in love with her.

It wasn't the end of the world. Spavins cause the hock joints to fuse and then are unlikely to be a great problem unless you have high aspirations, which I certainly didn't. She was given some anti-inflammatory powders and we had to follow a careful exercise programme to strengthen the muscles that would protect her from further problems. She thought this was fine. After all, there would be no school work for a while as riding in circles would be too much for her hocks in their present state. Instead we had to do lots of walking followed, after a few weeks, by a little trotting and then gradually bringing her back to her normal exercise routine. She was no bother at all and soon got used to the idea that we walked everywhere. When we did move up to being able to trot for a few minutes, it was really quite exciting!

A great highlight that spring was a trip to Gleneagles to see a demonstration by Monty Roberts. We were thrilled to get tickets for this sell-out occasion and decided to make a day of it, arriving at Gleneagles early so that we could watch all the preparations. It was truly magical, watching Monty work with a variety of horses, all presenting some form of challenge. What impressed me most as I watched him talking to people beforehand and then working in the round pen was his calmness. Nothing bothered him or seemed to surprise him. He just worked away quietly till he got the result he wanted. There was no shouting, no excitement, and no whips. There was just the simple message to the horse that life was pleasant if it was co-operative and hard work when it wasn't. Aidan was fascinated. He'd had a few challenging times attempting to catch Chelsea. Sometimes she was just being difficult but watching Monty, Aidan began to realise that, being so tall, he had probably appeared to square up to her and so (unwittingly) had been sending her away. He learned to soften his frame, slightly rounding his shoulders, and thus encourage her to come to him. That week, he was eager to put his new found knowledge into practice and since then, he has had few problems catching her. In fact, he's often the one who can persuade her to come and be caught even when she is playing hard to get.

All through the winter, things had become more difficult at the yard. From being a well run operation, it had gone downhill. The owner was away a lot and supplies for the horses often ran very low. Beds were sometimes very thin and hay was not always as plentiful as it should have been, nor was the quality as good. After Christmas, I found myself going up twice a day to look after Chelsea. It coincided with her hock problem and I really wanted her well cared for at this time. I especially wanted her to have a good thick bed of straw. I didn't want her slipping as she got up from lying down and I knew her stiffness might make this more likely. I took to mucking out and doing her bed myself. Nobody cared too much about what was happening and the staff seemed very glad to have one less stable to do. That way I could give her a really deep bed and if I was going to be away overnight, I knew she would be comfortable even if not much straw was added to replace what was removed when her box was mucked out.

Like most people, Chelsea is happier when she knows what to expect day to day. Gradually, there was less and less of a routine at the yard and care became a matter of convenience, rather than a response to the needs of the horses. Just like a stressed person, Chelsea lost weight and began to behave badly. She was increasingly reluctant to be caught and could be very grumpy at times. I had started to look for a new home for her but I needed to be sure that it would be a good place, especially when I was still away from time to time. I didn't want to move her from the frying pan to the fire. Fortunately, I was soon able to find a new yard, Loanhead Equestrian Centre, where conditions seemed to be first class with the thickest beds I had seen anywhere other than the National Stud at Newmarket. I arranged to move her.

What a transformation! She was already familiar with her new home as she had been there a couple of times to have her saddle checked. Even so her reaction to the new surroundings was amazing. I led her into an isolation box where she would have to stay for a few days. She spotted the automatic water and nudged it. I was pleased that she knew what it was as she had been given water in buckets at her old yard. She then made straight for the manger piled high with sweet hay. She sighed, visibly relaxed, and almost at once she was happily munching, knee deep in clean straw and looking totally at home. I hung around for a while, putting my tack away, sorting out her feeding requirements and passing on essential information to the staff. I checked her once more before going home and left her still contentedly munching. Already some of the stress seemed to have left her.

Next morning, I was delighted to find that she had been lying down during the night, always the sign of a relaxed horse. I took her out for a little walk in hand just to stretch her legs and familiarise her with her new surroundings. The following day saw the start of a major two day show at the yard. When I went to find her, she had already been moved to her permanent quarters in the main stable block to allow two stallions visiting for the show to have the more secure isolation boxes. The main block was buzzing with activity as horses came and went from their competition classes. Amidst the bustle, there was Chelsea enjoying a mid morning nap with her head down, comfortably resting a hind leg. What a change! Within a few weeks, her condition had improved dramatically and she was back to being her old self, happy and co-operative once more. The calm routine of the yard and the kindness of those around her worked as well for her as it would for any of us.

Chapter 6: *Settled Again*

Chelsea and I loved Loanhead. Although it was on a much bigger scale than we'd been used to, it was very friendly and the horses were treated as individuals. Many were competition horses, show jumpers and eventers mainly, and nobody worried too much about their quirks and peculiarities as most were good at their job. I didn't have to worry about being away from home as I knew she would be well cared for in my absence. Another piece of good fortune was that one of the girls who had been at the old yard was now working in town for one of the oil companies and was delighted to have the chance to exercise Chelsea for me when I was away. Chelsea loved Kirsty so both of them were happy with the arrangement.

We were still following her fitness programme after the episode with her hocks but she was now working well, able to trot for longer periods and also to do a little school work. I began to think about some competitions. Our main aim had always been endurance and so I began to work towards a couple of competitions later in the season.

In the meantime, I received an entry form for the North East Horse Show which was to be held at Loanhead early in July. It looked like fun; it was a home event and thus was irresistible. I read through the classes and decided to show her in-hand in the adult hunter class. I didn't feel her school work was advanced enough for the ridden classes nor did I have much clue as to how she would react to the atmosphere of a competition so altogether the in-hand class would be a good place to start. In any case Chelsea is very pretty so I hoped she'd catch the judge's eye. I started to do a bit of practising so that she would walk and trot well. She was good to lead so it wasn't difficult, more a case of simply sharpening her up. We also worked on standing straight so that she didn't look quite as pigeon toed as she actually was. Normally I didn't mind about that as being 'in' in front and 'out' behind is often the mark of a good jumper but I thought we might as well look our best for the judge.

I had to get myself sorted out too so that I looked the part. I found I needed a whole new wardrobe. I acquired a hacking jacket, show shirt, tie, beige jodhpurs and short, black boots along with gloves and a show cane. It all felt very strange after my usual, casual riding dress.

During the week before the show, Chelsea had her mane and tail pulled and her feathers trimmed. The day before, she had a bath using 'Fiery Chestnut' shampoo which left her glistening like copper in the sun. I was up at crack of dawn on show morning. Aidan was golfing so he wasn't able to help but I had a couple of friends lined up for expert assistance and moral support. The main task was to plait her mane. I had been doing a bit of practising but was glad of some help. I brushed her till she shone and put some oil on her hooves. Then there was the old show trick of rubbing some baby oil round her eyes to make them stand out. On with her bridle and she was ready, looking really stunning. Her beautiful Thoroughbred head looked alert and ready. I did a quick change myself and we were all set for the judge's scrutiny.

From the moment we walked out of her box, she knew she was the queen. She walked regally over to the show ring and waited quietly until we were called in for her class. I have no idea whether she'd ever been shown in-hand or under saddle but she certainly knew about presence in the ring. She made me smile. It was a hot day and there were lots of flies. We'd applied fly repellent to her but they were still annoying her. As we stood waiting outside the ring, she had shaken herself vigorously. Once inside, however, she limited herself to a gentle, discreet shake just as she does if I am riding her. She definitely knew this was serious!

We walked and trotted in a big circle for the judge. As we passed him, I remembered to smile and Chelsea turned her head to smile too! Imagine my amazement when we were called in 2nd after the initial inspection. Although it was a fairly small, local show there were some well bred horses in our class and we were standing above them. I was under no illusions, however, and I knew that when the judge spotted her rather arthritic hocks, we would be moved down the line. Still, it was pleasing to know we'd made a good first impression. Soon it was our turn for an individual inspection. Chelsea tolerated the judge feeling over her with good grace. We then walked and trotted for him and took our place in the line once more. Chelsea behaved impeccably, standing quietly even when another horse in our class broke loose and galloped out of the ring. Soon we were walking round for the final time. This time we were called in 4th. I still couldn't believe it. It was more than I'd dared to hope and we had certainly outclassed many of the well bred horses. Clearly the judge liked good, old fashioned Irish hunters such as Chelsea and of course I have to agree with his preference. This was our first ever competition together and the very beautiful rosette still has pride of place in our collection.

A couple of weeks later, we did our first endurance ride together. This too was great fun and a much more serious test of her hocks. It was also our first 'away' competition and a chance to test out our new lorry. Everything went smoothly. We passed the vet and the tack inspection and were soon on our way. Last time Chelsea had been at this location, she had been doing cross country. The first part of our route passed some of the fences, so I had to take a good hold as it certainly crossed her mind that the quickest way through was to jump them!

We soon settled into a steady rhythm and headed towards the hills. I had been a little anxious about how she might react to faster horses coming up behind, overtaking us and going ahead, as she can be very competitive and I didn't want her to start racing. To my great surprise and relief she was as good as gold. Her ears told me when horses were approaching and she would obediently stand to one side to let them pass.

Our time was a bit slow as I had been reluctant to push her and Aidan had met us at several points along the way just to check that all was well. I still wasn't sure how her hocks would cope. In fact I needn't have worried. The time didn't really matter. We'd had great fun and she'd completed the course in great shape, her heart rate at the end being lower than when she started, showing how relaxed she was at the finish. Next morning I trotted her up and was delighted to find that she was as sound as a bell. I turned her out in the field for a well-earned day off.

Our next outing was a twelve mile charity ride on the beach to the north of Aberdeen. This was great fun, after a slightly sticky start. I had chosen to ride with a 'mostly trotting' group which was made up of a mixture of horses and ponies. Our leader was a ferocious Pony Club District Commissioner who insisted that we must stay behind her "at all times". Seeing Chelsea jog along in anticipation of things to come, she insisted we ride up close behind her, presumably to keep an eye on us! This was really just as well for as we got down to the beach, her horse baulked at the narrow path over the shoulder of the sand dunes. Not even a tap with her whip would encourage it to go forward. Clearly the horse was less afraid of her than we were! The second horse in the line refused to go too, having seen the other one refuse. "Chelsea will lead," I found myself saying and the other horses gratefully moved aside. I urged Chelsea on with an encouraging pat and a gentle squeeze with my legs. The path was narrow and awkward but brave creature that she is she plunged into the soft sand and on across the sand dunes down to the firmer sand on the beach. She also led our little group over the rivulets that ran across the beach.

By then she was raring to go. It wasn't that she wanted to gallop off. She was just fed up having to 'amble' along at the speed of the ponies. She was trying to be good but wasn't happy. I decided to risk the wrath of our leader and asked her if we might go ahead at our own pace. Having had me sign a verbal disclaimer to absolve her of any possible consequences, she allowed us to go. We both gave a sigh of relief and trotted on towards Newburgh. Chelsea was really happy now, able to cover the ground easily with her flowing trot, interspersed with the occasional canter. Aidan and a couple of friends were waiting for us at Newburgh where we had a quick cup of tea with a couple of apples for Chelsea. We didn't hang around as I didn't want her to get stiff standing in the coolish wind. When I explained about Chelsea's hocks, we were granted permission to ride back on our own. That was just the best - my lovely horse and six more miles of glorious sunny beach.

It didn't take us long to get back. In fact Aidan only just got there quicker by car! I had noted the point where we had first come down on to the beach so that we could find our way back. I needn't have worried, though. Chelsea had also taken note and spotted the exit over the sand dunes before me. Aidan met us and we walked up the last stretch of track together.

Back at the start, we handed in our number and collected our rosette. It had been great fun.

We had one more endurance ride to finish the season in October. This time I was prepared to be more competitive and so we pressed on at a brisker pace. It suited Chelsea much better too and I definitely felt she was my willing team mate. We finished in a good time despite Chelsea losing a back shoe some six miles from home. I was prepared to lead her back but she had other ideas and clearly wanted to end the season well. I decided it was less stressful just to let her go at her pace and we had lots of canters on the lovely soft, grassy forest tracks which she loved. She trotted up sound for the vet and again her low heart rate confirmed her fitness and also showed that the shoeless foot was not causing her any discomfort.

Chapter 7: *Thinking Time*

Aidan has had an interesting and successful career but has often found himself in difficult circumstances. He has been made redundant and then gone on to find himself a survivor of massive world-wide job cuts. He has seen good times and bad. Over the course of the summer, he began to learn about the moves to bring Spirit and Values to the workplace from one of his golfing partners who had attended a programme run by Lance Secretan in Canada. It struck a chord with him and he wanted to know more so he enrolled on a retreat at Findhorn. I have to say that it all sounded a bit 'touchy-feely' for me. I couldn't actually go to the retreat anyway as I had already planned to do a riding course in Wales at the same time. Nevertheless, there was something interesting about it.

Aidan was enthused and inspired by his experience and the company which organised the retreat liked him too. He started to do some further training and shadowing with them. Gradually, I became more intrigued too and we both decided that we wanted to attend the annual Secretan Centre Retreat the following summer. I began to realise that my own business ethics were not so different so maybe it wasn't so 'way out' after all...

One of my long time management gurus is Charles Handy. I had the pleasure of meeting him whilst doing a year's course at what was then Oxford Polytechnic in the early 1980's. He presented an interesting and, as it turned out, very accurate picture of the world of work in the future. I had read all his books and found myself re-reading "The Hungry Spirit" which referenced books by Lance Secretan and so I read these too. I found much to relate to, reminding me that the approaches that had worked so well with Chelsea could be a bridge for people too.

Back at home, the riding course had been a great success and I was looking forward to trying out my new skills with Chelsea. I had spent the week concentrating on show jumping so at the first opportunity I booked up to do a grid session with her. Gridwork is basic training for show jumpers and I thought it would be a good place for us to start again. After a somewhat uncertain start, we found our stride and had a great time. I'd forgotten how much fun she was to jump and she loved popping down through the grid. I began to make more plans for the Christmas break.

It was as well we had that session for suddenly black clouds began to gather. Sadly Aidan's Dad died just before Christmas. He had been frail for a while but he always seemed to recover from setbacks and somehow we had come to think of him as immortal and so it was a shock when he passed away. Christmas and the Millennium celebrations were muted for us and we were glad to get away for a week's break down to Fife in the New Year. With all the upheaval at home, I had given Chelsea a break over Christmas and planned to get back to work with her when we got back from holiday.

As ever, I was looking forward to our first ride out together after we got back. We had a walk out in the woods, but it was soon clear that all was not well. She was jogging, throwing her head around and was really not herself. When we got home, I had a good look at her. I couldn't see anything obvious in terms of heat or swelling anywhere. I decided to get her saddle checked as it was due to be done. Next day, I looked at her again and found a warm spot on her back. She was still not very comfy so I decided to have the vet check her out. She was due for a flu jab in any case so it would certainly not be a wasted visit.

The vet arrived to check her over and I was glad I had called her for, by then, Chelsea was showing signs of lameness in her right hind leg. The vet checked her over and thought that the most likely cause was the spavin in that hock not fusing. However, it would take X-rays and other tests to determine this conclusively. Chelsea had been using her back to save her leg – hence the discomfort there. I felt so sorry for her. We were able to arrange for her trip to the Animal Hospital for the following week and in the meantime, she was left to rest.

We were very fortunate to have such an excellent facility more of less round the corner from the yard so she didn't even have to endure a long journey. She was an angel throughout the three hours of tests. Despite being lame, she thought she was going to an endurance competition, an idea reinforced by seeing another horse being trotted up for the vet and so was very much on her toes to start with. However, she gradually got the idea that it wasn't a competition and resigned herself to the various tests and hanging around in between them. She didn't even mind the X-rays. She walked into the X-ray suite and stood quietly whilst all manner of machines were placed round her back legs.

All the tests confirmed that the spavin in her right hock had not fused naturally. Fortunately the one on her left hock had. Given that the first level treatment hadn't worked, the vet recommended that she should have a steroid injection in her right hock to encourage the fusion to take place. I was happy to accept her recommendation and the treatment was planned for the following week.

A steroid injection is a painful process for animals as it is for people and so Chelsea had to be heavily sedated for the treatment to take place. The injection had also to be placed very precisely which is another reason for having a very still patient. On the morning of the injection, the stable staff gave her a very deep bed, well banked up round the walls so that if she lay down afterwards and found it harder than usual to get up, she would come to no harm. The vet arrived and gave Chelsea a friendly pat. Given her past experience with Lesley, I had every expectation that Chelsea would associate her with needles for flu jabs and tests for her spavin. However, she bore her no grudges. She knows that Lesley, herself a keen horsewoman, really does love her patients and she often has polos in her pocket! We stood Chelsea diagonally across her box so that there was plenty of room to work behind her. She had a pre-med and then the full sedative. She remained on her feet, as intended, but her head sank downwards and it was clear that she was no longer aware of us.

The vet quickly found the spot for the injection and the treatment was soon completed. Lesley expected the sedative to last for about an hour. I wanted to stay with Chelsea so I settled myself in the corner of her box, leaning out over the half door whilst watching her and talking to her. It is always said that hearing is the last sense to go so I thought she might like to hear a familiar voice. I was reluctant to touch her as I didn't want to startle her, while she regained full consciousness. I remembered how jumpy our cats can become as they emerge from sedation. Gradually she came to, looking just as dazed as any person coming round from an anaesthetic. I asked her how she felt and told her it was all over. Head up and eyes more alert now, she turned to me and rather gingerly came over to me and nuzzled my hand. I stroked her gently and told her she would be fine. Then she turned round and started munching hay. I was immensely touched by this gesture. I like to think she had heard my voice throughout and had now wanted to reassure me that all was well.

With the injection behind us, the next stage was to get her working again. At the vet's suggestion, I tried to ride her. It was terrible, just like sitting on a rickety table and I was close to tears. Would she ever be sound again? Lesley didn't seem too concerned and suggested

I lunge her in very wide circles instead until she was less stiff. She also arranged for Chelsea to see a physiotherapist.

The physio was a kind lady but for some reason, known only to herself, Chelsea didn't take to her. This made it difficult for the physio to carry out an effective assessment on her so I had to do the tests under her instruction. We then found that, as long as I had my hand on her neck, Chelsea would tolerate the physio's touch. The main treatment recommended was the daily use of an electro-magnetic pulse rug. I was slightly anxious about using this as it involved Chelsea being plugged into the mains electricity via the rug for twenty minutes at a time. I wondered whether she would stand still for so long. I needn't have worried, though. It was obviously blissful treatment and once I put the rug on and plugged it in she dozed very happily.

The lungeing was hard for her to do even although I walked about to create very large circles. She was very stiff but she just had to work through it and gradually began to move more easily. I was encouraged as the physio, who while watching one of our sessions, remarked that she was a good athlete and had every chance of recovering well. I also loose schooled her and was interested to watch how she moved when left to herself. Not surprisingly, she tried to avoid using her right hind leg whenever possible. Although the steroid injection provided pain relief as well as encouraging fusion, I think she was anxious about hurting herself, and yet was still keen to go. It was a balancing act -

I just needed to encourage her to have more confidence in her weaker leg.

Soon I could ride her again and although she still felt stiff, we enjoyed walking out in the woods. As her movement improved, it was time to try trotting again. She was fine on her left side, but when I asked her to use her right side I could feel the stiffness again. I persisted gently to see whether she would be able to work through it or if it would get worse. I could see her thinking about it, her 'quarter to three' ears telling me she was concentrating hard. After a few stiff strides, I could feel her stretching and started to step out more confidently. "Good girl," I whispered and gave her a pat. After a few good steps, I let her come back to walk and I gave her another pat. Then I asked her to trot again on her right side and this time she was much better. Again after a few strides I jumped off and we walked back to the yard together. I wanted her to know that I was grateful for her willingness to try. It was just so typical of her – honest, tough and really wanting to get better.

Chelsea's favourite pace is gallop or, at least, a good going canter. I'd always known this but during her rehabilitation it became clear that it was really very dear to her indeed. I was careful to increase her exercise very gradually. We had got to the stage where we were still doing plenty of walk and a bit of trotting but canter was still a few weeks away. For no reason, however, she became quite stuffy. There was no obvious problem but she didn't seem to be her usual enthusiastic self. Fortunately this coincided with a check up from the vet and I explained what was happening. Lesley watched me ride and agreed that there was no obvious problem but that something was definitely amiss. She then rode Chelsea. What a joy to have a vet who is also a good rider! Again she was sure that there was nothing actually wrong. Then she had an idea. "What would happen if I asked her to canter? Will she buck?" I explained that she would only buck if she was sore but added that Lesley would certainly know when Chelsea moved up to canter! She's short coupled and very bouncy, like a hard suspension sports car, so as she strikes off her hind legs into canter, you certainly know about it. The vet trotted round and asked for canter. Wow! Chelsea's response was enthusiastic to say the least and she bounded off round the school, grinning from ear to ear. That was the moment we decided to tear up at least part of the rehabilitation programme. Chelsea certainly hadn't read the section that said "no cantering yet". This didn't mean that we galloped everywhere, but a couple of times a week on good going, we had a little canter. It worked a treat and helped her cope with the steadier work that was essential to rebuilding her strength. I must admit that I enjoyed these little bursts too!

As with trot, she was reluctant to use her right hind leg to strike off in canter. I could sympathise with that, having had an ankle injury myself many years ago. It was a while before I would trust it enough as a take-off leg whilst long jumping. I decided to see if I could to lull Chelsea into using her right leg. One of the canter tracks she loved was quite twisty and she always did wonderful flying changes to stay balanced as we cantered round the bends on our way up the hill. If we started on her strong leg would she just change anyway as she had always done? Sure enough, it worked perfectly. I slightly adjusted my weight as we cantered round the bend and she changed to her right leg to stay balanced. She was having such a good time that she had forgotten about which leg she was using!

Now she was going so well and her balance was returning to normal once more, it was tempting to start competing again. However, I decided not to. She was using herself better than ever before but was finding it hard to sustain so I decided that more fitness work would be time well spent. I also knew that developing all the muscles correctly would help to prevent problems in future.

Although in some ways this was not very exciting, it was full of surprises. Most of them were pleasant – her beautiful extended trot through mud, her ability to see her stride over rough ground, her joy at finding anything that could be turned into an athletic move such as the mound that became an Irish bank and the sharp corner at the bottom of a hill that could be cut with a single bound to provide a launch off into canter up the hill on the other side.

The first year of the new millennium was certainly proving a time of change for us. Aidan had started on a project in Holland in January just after our holiday in Fife. It was supposed to be for six weeks whilst the team settled in but then it stretched to Easter, then June, then

September and finally until Christmas. It was hard going, being away so much. Each Monday, he was up at 4.00am and away to catch the 6.20am flight to Amsterdam, returning back into Aberdeen at 3.00pm on a Friday. We began to live for the weekends.

To add to the challenge, Aidan's sister came to stay for six months. She had lived in South Africa for many years and had found it hard to settle back in the UK, especially in London which had changed so much since last she'd worked there some 25 years earlier and so she had decided to come back to Ballater and look for work locally.

With all this going on at home, I had decided to cut back a little on work. In the light of Chelsea's hock problems this was especially welcome. Furthermore, I was to be glad of this easier workload as it gave me time to do some thinking. In some ways, I suppose, it made up for having cut short my planned long break two years before.

As Aidan's sojourn in Holland stretched out, he was able to negotiate about a week every month at home. He spent this time with the people who had run the Retreat, shadowing their work and learning more about their approach. I learned too and we began to look forward to the Secretan Higher Ground Community Retreat to be held in July in Canada.

We decided to make a holiday out of it and planned a few days in the Ontario countryside near where the Retreat was being held beforehand and a couple of days in Toronto afterwards. We fell in love with Canada and thoroughly enjoyed our holiday, especially the opportunity to try riding Western style in the Blue Mountains. Aidan loved this too and felt very secure in the large comfy saddle. I had an amusing moment whilst waiting to go out for our ride. I was standing by the paddock talking to one of the horses when I was approached by an American family. "Gee," they said, eyeing my riding gear, "You look like you know what you're doing. Say, how do pat a horse?" I had to think about that for a moment. I just do it instinctively and these things are always the hardest to explain. Anyway, I suggested that it was best to start on the neck and let the horse get to know you a little. Soon they were making friends with one of the kind trail horses.

We duly made our way to Barrie on the Thursday afternoon and followed the instructions to the Kempenfelt Centre where the Retreat was to be held. It was a beautiful spot on the edge of Lake Simcoe. We registered, got unpacked and went to explore. We met up with the people we knew from home, Aidan introduced me to others he had met at the Retreat in Findhorn and we put faces to those whom we'd only spoken to by e-mail and the Web. All around others were reconnecting with friends not seen in a while with shrieks of joy and huge hugs.

Everyone was kind and friendly and we soon felt part of the event. The Retreat proper started on Friday afternoon and soon we were sucked into the round of sessions, some involving everyone and others being smaller break-out groups. Aidan and I had deliberately chosen to go our separate ways for these so that we could learn even more by exchanging our experiences. At times I felt it was all a little strange and some of the sessions floated over my head completely but the overall theme was one of music so I never felt completely at a loss. The energy generated was quite amazing and everyone was so friendly that I soon began to feel more at home. I didn't feel uncomfortable when things floated past as I knew that I would eventually understand. I also knew that I needed to slow down a bit as I struggled with some of the more tranquil, meditative sessions. There was a lot about 'Living the moment' and I'm usually a few leaps ahead!

We came home exhilarated and refreshed, determined to attend the Secretan Associates Leadership programme in November. We had felt at home with this family and wanted to belong to it more formally. Aidan went off back to Holland and I used August, a quiet month work wise, to do some thinking and reading.

My back had never quite recovered from the slipped disk and often when riding I felt very lopsided. For some time, I had been considering taking lessons in the Alexander technique, a gentle discipline that improves posture and movement. I pulled out a book about the Alexander technique and riding which I had had for some time. Typically, on the first read through, I had gone straight into the really practical bits about exercises in walk, trot and canter. This time, I was a little more thoughtful in my approach. Imagine my joy at finding a chapter about 'Living the moment', a tenet held dear to Alexander followers. In this overall context of riding, it made so much more sense. Another reason to be grateful to Chelsea as it was my desire to be as balanced as possible on her back that had brought me to this discipline. I sought out a teacher and was delighted to find that there was one living not too far away.

My Fridays took on a new routine – quick tidy up at work, an Alexander session, the weekly supermarket shop and off to the airport to meet Aidan. Together we'd visit Chelsea and then set off home to enjoy the weekend. Initially I had lessons several times a week to make the habits more permanent. Interestingly, my teacher found that to start with, she had to help me move quickly as I struggled if she went too slowly! Our goal was to help me slow down my physical movements – and gradually it began to happen. I found that the position suggested for relaxation was also good for meditation and so my journey of discovery moved on. There was a lot to do and to read before the Associate programme in November and late summer quickly melted into autumn and soon we were counting the days till we went away.

Meanwhile, Chelsea had fully recovered and was going well. We'd missed all the summer competition season but it didn't matter as giving Chelsea that bit of extra time had been so worthwhile. Since moving to Loanhead, I had been having lessons with Jackie Tague (now Mather) and she also rode Chelsea a few times when I was away. They got on well together as Jackie was kind and sensitive with Chelsea, whilst at the same time quite firm when necessary. Sadly, Jackie's main competition horse had broken a foreleg during a wild gallop out in the field and, although expected to make a full recovery, was going to be out of work for a while. She had been looking forward to doing the winter dressage series with him and so I asked her if she would like to compete with Chelsea instead. She was delighted at the opportunity. Chelsea wasn't quite so enthusiastic as the blind obedience required for dressage is just not her cup of tea. However, she does like the chance to show off and is always up for any kind of competition.

The first competition was out of doors and much to our surprise she won her class and came fourth in the qualifier. She and Jackie made a wonderfully elegant pair and, for once, Chelsea did as asked for each of the required movements. Encouraged by this, they continued to compete all winter, going on to have several more wins and being placed on the other occasions, finally qualifying for both Prelim and Novice finals. Sadly, they never got to compete as the finals were cancelled due to the Foot and Mouth outbreak. More pleasing than the rosettes, however, were the judges' positive comments about her way of going. Although she was occasionally tense as she tried so hard to be totally obedient, there was never any mention of stiffness.

Before going to Canada for the Associates Programme, we managed to attend the last day and a half of another retreat at Findhorn. I was now feeling more at home at such events and it was good to meet old friends and make new ones. At the closing session, I was touched that so many people used the word 'Gentleness' about me. I had been trying to be a more gentle person or perhaps I was just confident enough to let my gentleness shine through more. Again, when we came home, I was glad to have more time for reflection.

Around this time, I was reminded at how good animals are at tuning into a world we humans can't perceive. Late one afternoon, I decided to take Chelsea out for a quick ride round the woods. It was October and darkness was not far away. The forest was quite atmospheric as the sun shone through storm clouds between heavy showers. We're quite used to being out in all weathers but Chelsea was clearly worried by something. She became reluctant to go forward which was very unlike her and eventually was almost shaking with fright. I couldn't see or hear anything unusual. I decided to get off and lead her for a while as this usually calms her. I was also a bit worried as we were on a narrow track though heavy forest so had she bolted my chances of steering her safely between the tree trunks and branches were almost zero.

She was better with me beside her but still upset, nuzzling my hand and glad of a reassuring pat. Quite suddenly and for no apparent she settled and I got back on board. It was almost dark now, but her anxiety had passed and she was back to her old self, trotting and cantering amongst the autumn leaves as we headed for home. I was still puzzled about her strange behaviour.

Imagine how I felt when I read in our local newspaper the following day that whilst we had been out there had been a serious road accident near where we were riding. Sadly it had claimed the life of a young policeman and had injured several other people. I had just heard distant sirens - but that was not unusual on the busy roads that run near the forest. Chelsea must have heard a lot more. Did she think she was carrying us into danger? Was she worried by sounds of people in distress? Out of curiosity, I decided to look at a map. To my amazement, all the time that Chelsea had been anxious we were going to towards the site of the accident, as the crow flies. The point at which she calmed down was where the track turned away from the accident. Not for the first time, I was glad I'd listened to her and respected her feelings.

The Secretan Associate Programme turned out to be one of these rare and totally life changing experiences. It didn't start very auspiciously for me. I must have picked up some tummy bug on the journey and as a result was sick for most of the night before the programme started. I was still feeling very poorly in the morning. However, not only was I very comfortable in our lovely room in the old farmhouse at Kingview, but I was surrounded by love and concern too. I was left to rest, but not before I had been introduced to the miracle of attunement which left me feeling very relaxed and able to catch up with some sleep. From time to time, Aidan came by to tell me what had been happening on the course. One of the things he wanted was "something of significance" to place on the altar for the duration of the course. I had a quick think and gave him the photo of Chelsea I carry in my wallet. Little did I realise how much she was to play a part in proceedings. Throughout the day, I felt included in the happenings with the energy of the group reaching out to me constantly.

The next morning I felt much better but still a bit fragile. I asked for a moment at the start of the day to say my thanks to everyone for their kindness. However, everyone wanted me to feel a real part of the group and so I had a short welcome ceremony all to myself. The previous day I'd been given a small crystal by one of the course members whom I'd met in the summer so I added that to the altar. Everyone smiled when I said that what I'd left behind to attend the course was Chelsea. My love for her was already well-known.

The week was challenging, interesting but most of all enjoyable, spent in the company of a wonderful bunch of very talented and gifted individuals. At times I ran out of energy but at other times felt I'd moved on a lot from the summer and the person who'd attended that last Retreat. I was especially aware of this when Lance did the 'Calling Meditation'. This was one of things that had floated by me in the summer but this time it meant a lot. By now, I had decided that my calling really did lie in bringing together my consultancy and coaching skills with my love of horses. It had always seemed too complicated before and now it was so obvious. Aidan was delighted to share this future which added to my joy. The word gentleness had cropped up again too so it was clearly meant to be part of our future.

For Aidan, this offered the possibility of book writing – a long held dream for him. For me, it was bringing the horses into my work. We had a new way forward.

Chapter 8: Transition Time

We returned home enthused and determined to press on with our new plans. Aidan had to finish in Holland first so it was another few weeks of commuting over the North Sea and there was plenty to do in London for me. Finally, Christmas arrived and the chance to relax and reflect at home. I had been saying to Aidan for some time that he should not rush to find work in the new year but should take some time out to unwind. It had actually been about eighteen months since he had lived at home for any length of time as, before the Holland project, he had spent many weeks in Chertsey, just outside London, on another project.

Writing "Gentleness" the book was just the sort of project he needed. It was absorbing but at the same time unpressured and could be done from home. In the meantime, I carried on with my project in London but there was always time for reflection too.

I was glad we were not travelling so much as 2001 started with the heaviest snow here in North East Scotland for many winters. Snowy weather doesn't worry Chelsea at all and on more than one occasion, we got caught out in a snow storm and came back looking like the Abominable Snowman on Horseback. Often, I have been inclined to turn back but Chelsea was the one who wanted to keep going and so we did, always having great fun.

During one of these snowy outings, on a quiet, still Sunday morning, Chelsea came to an abrupt halt. I couldn't see anything unusual, so I patted her and nudged her forward, but she wouldn't move. All I could see was a man coming towards us on a nearby footpath with just his head and shoulders showing above the bushes. Nothing unusual about that I thought and so I asked Chelsea to walk on, a little more firmly this time. Still she refused to budge. As far as she was concerned, this man had apparently arrived straight from Mars and she was going nowhere near him! After another look, I understood. She was quite right! He might well have been from Mars, for he was wearing cross country skis. Their quiet shoosh on the snow and the man's gliding motion, both of which she'd picked up long before me, had been enough to alert the ever observant Chelsea. I then engaged a ploy I've successfully used to reassure her many times around 'unusual' people. I called out a cheerful "Good Morning" to him and, fortunately, he replied in equally cheerful terms. That was enough to reassure Chelsea that he was perhaps from this planet after all and she was happy to respond to my request to walk on.

By mid February, we were tiring of the snow as the tracks had become very icy and difficult, really too dangerous for riding. To make matters worse, Foot and Mouth arrived in UK. Although horses cannot contract this terrible disease, they can carry it, just as humans can, and so my lovely Chelsea found herself confined to the yard. Initially, with the poor ground conditions, this didn't matter too much and we were blessed with good indoor facilities. In any case, we felt for all those farmers whose livestock and livelihoods were at risk.

As February gave way to March, though, Chelsea was definitely getting bored and became grumpy and rather naughty. To add to her woes, she developed an abscess in a hoof and had to be confined to her stable whilst it drained and healed. As ever, she was an angel to nurse, standing quietly twice a day with her foot in a bucket of water to clean it out then holding it still whilst I bandaged it.

We had planned a week's holiday but went off with an easy mind, knowing she would be well cared for in our absence. Imagine my joy at returning home to find that she was completely fit again. Even better, the forest tracks were open once more, the Foot and Mouth restrictions having been eased. Next morning there was no stopping me. A bit jetlagged, I was still up at 5.30am and riding at 6.30am enjoying the first early morning ride of the year, almost a month later than usual. Chelsea was delighted to see me and realised that she was about to be released from her comfortable confinement. As I led her out of the stable, passed the entrance to the outdoor school and over the road to the forest, I could feel her anticipation - "Yesssssssss....."

I had planned on a gentle walk as I thought she might be a bit stiff from her fortnight of confinement but I soon realised that on this glorious morning of sunshine, singing birds and scampering squirrels, Chelsea had celebration on her mind. She certainly wasn't stiff, her swinging steps at walk told me that and I could feel all the pent up energy ready to burst forth.

We rode round the edge of the forest under a canopy of still bare branches, enjoying the sunlight and the breezy wind. I abandoned the idea of a sensible walk and with the lightest nudge of my legs, Chelsea exploded into her usual energetic canter up to the farmyard, dancing in the wind and playfully spooking at the shadows swaying in the breeze.

The farmyard was just a few buildings which, at that time of year, were the maternity unit for a small flock of sheep. The lambs had arrived and were enjoying a frolic in the sunshine. Their mothers summoned them as they heard us approach. I hoped they would be safe from Foot and Mouth.

Along the track, we met some deer. Chelsea spotted them first as she always does. They were enjoying the shoots of new grass. I wondered if they were the same little group who had warned us of the hard weather when we saw them in January, making their way deep into the shelter of the forest. They had wintered well and I wondered if we might see some babies come June.

Chelsea was delighted as we turned up towards the gallops, well named and one of her favourite stretches of track. I didn't have to bother with a nudge this time. "Let's go" I whispered to her and leaned forward in the saddle. She was off, bounding up the hill like a deer, mane flying in the wind and her chestnut coat gleaming in the sunshine. We paused at the top and walked down the other side of the hill. At the bottom she cut the corner and galloped up around the other side of the hill. It was more sheltered here and so there was a green fuzz of early spring growth on the silver birches.

We watched the commuters making their way to work in the city and I hoped they'd take a moment to enjoy the glory of the early morning, to carry a little of it to their desks and to their colleagues. As we ambled home, both thinking of breakfast, I gave Chelsea a grateful pat for sharing her joy at being out and about once more, for spotting the deer and for helping me appreciate the beauty and joy of a spring morning.

With all the Foot and Mouth restrictions, competition prospects looked bleak for the summer. I had missed the build up to the endurance season, but kept Chelsea ticking over anyway. Fortunately, Chelsea had had a busy winter so a quiet start to the summer was not going to be too great a disaster.

As well as taking part in the dressage competitions, Chelsea had begun jumping again. I knew how much she loved to jump and so, as her fitness improved, I suggested to Jackie that she might try her over a little jump. Jackie needed no second bidding. I set up a cross pole with a canter pole in front just to help Chelsea set her stride. I could see hope and excitement in her eyes as I built the jump. "Is that for me?" she seemed to be asking. Jackie trotted her round to the jump and she knew at once it was indeed for her. Her ears locked on to it and she set her stride to clear it easily. Round they came again and Chelsea showed her contempt for its height by clearing the canter pole as well in a single, easy stride. Fortunately Jackie had the skill to sit tight! Gradually, we introduced more jumping into her programme – much to her great joy. During the winter, we decided that Jackie should take Chelsea round a full course of jumps at one of the fortnightly Wednesday shows. Chelsea thought this was a great idea. Although very green after her long layoff from jumping – it was about two years since she'd jumped a full course – she attacked the course with her usual gusto. Jackie described the experience as being "taken on a guided tour of the jumps"! It was a clear round first time out. The jump off was just as exciting but even quicker with Chelsea turning for the jumps even faster. They ended up a very creditable 5th. Wednesday evening jumping became a regular outing which we all enjoyed and gradually Jackie managed to persuade Chelsea to listen to her at least occasionally and to slow up just a little. Even so, their rounds remained highly popular with spectators!

Initially she was a bit ring-rusty, but she always managed to jump even from a less than perfect stride. Jackie was content to sit tight and let her sort herself out – tactics Chelsea appreciated. I loved to watch her as she waited for her turn to jump, standing ring-side completely focused on the horse before her in the ring. I came to the conclusion that she was actually memorising the course. Nothing disturbed her concentration in these moments and her expression always reminded me of world class human sprinters as they stare down the track before their race.

Eventually as the Foot and Mouth restrictions eased in Scotland and outdoor jumping resumed. The Wednesday shows started again but this time out of doors. Chelsea loved this as there was more room to canter between jumps! As with the dressage competitions, I used to give her a good work out in the woods by way of preparation although in this case I'd do it just beforehand to take the edge off her energy. We were never sure whether this really worked as sometimes it felt more as if she'd had a thoroughly good warm up, to give her even more energy for the real event! Still, she and I enjoyed it greatly.

2001 proved to be a blissful year with Aidan working mostly at home and my visits to London being generally limited to just one day a week. Consequently Aidan spent more time coming out with us as we enjoyed early morning hacks in the summer. He would walk with us for some of the time, other times taking one of the 'people tracks' whilst we had a trot or a canter then we'd all meet up again and walk together for a while. It was a lovely routine which we all enjoyed not least Chelsea.

Chapter 9: *New Partnerships*

Ironically, it was Chelsea's being slightly lame that led me to buying Susie although I had made it known that I was on the lookout for another horse. I knew that Chelsea would always be limited by her hocks and I was also thinking ahead to my plans to bring horses into my work, even though I still didn't know exactly what form this would take. That afternoon, Chelsea had been badly in need of shoeing and afterwards was a little sore. I'd planned a lesson and so felt rather disappointed at having nothing to ride. Callum, who runs Loanhead along with his partner Shona, suggested I try Susie whom Shona had recently bought to sell on. Just as with Chelsea, it was love at first sight and I decided she was the perfect 50th birthday present.

Chelsea and Susie are quite different characters, each clever and lovable in very different ways. Chelsea has very sharp, typical Thoroughbred intelligence which she loves using. She is always right up there with me, suggesting what we should do and determined to go at every opportunity. In total contrast to Chelsea, Susie is a very sweet natured, straight forward mare. She is also very responsive and bright but has no desire to be the boss. She loves helping and supporting me in every venture, trying so hard to please.

Susie was only five when I got her and had had very little education so I needed help to bring her on. Jackie was about to leave to live in USA and so she recommended Sue Hendry, an excellent equestrian sports coach whose focus is on building successful partnerships between horse and rider. She was exactly the right person to help with Susie. During lessons, I inevitably talked about my plans to work with horses. Sue thought it was a little strange at first, but she was also curious. We had coaching as a common interest and that gave us a basis for discussion.

To start with, Susie was too thin and unfit to work for more than about 20 minutes so school work came along very slowly but she gradually put weight on and became stronger with long days out in the summer sunshine, good grass and two meals a day. Initially I just wanted her to show me what she knew. In Susie's case, this was simple. Leg on meant walk, more leg meant canter or canter faster. There was no in between - trotting was a complete mystery to her! Pulling on the reins got the faster response too as she'd never learned to take a steady contact. This latter 'feature' was good in one way in that she had a wonderfully soft, unspoiled mouth and she did stop if I simply took my leg off! She must have been used to her rider being legged up or vaulting on for she had no idea that she was meant to stand still whilst I mounted.

There was no point in being angry with her or telling her she was stupid when her reactions were not what I expected. I simply had to show her tactfully that I wanted something a little different. So, we worked away steadily on the idea that between walk and canter came trot.

She soon began to get the hang of it – initially taking a few canter steps then settling back into trot but gradually working out that she could actually go straight from walk to trot! She also must have learned to jump when out hunting as she had no idea that she could trot over poles. Our early attempts at this were most amusing until she worked out how to co-ordinate her legs!

In many ways, she was like some of the community groups I was working with in London at the time. They had come to a different culture just like Susie. It was important to recognise that they were not being stupid or ignorant when they behaved in seemingly inappropriate ways. It was simply that they did not yet understand what was expected of them in their new situation. It was also important to recognise the new and different skills they brought with them.

Her jumping ability was without question and it wasn't long before I tested that ability for myself. Having had a few pops over little jumps out in the field, I decided to try the smallest class at a Wednesday show. It was nerve-wracking but also hugely emotional. Whilst I was trotting round waiting for the bell to begin my round, the tannoy announced "First to go is Beth Duff and she rides Sparkling Rose" (Sparkling Rose is Susie's show name). I had a big lump in my throat for I had only ever heard this before in my childhood dreams. It wasn't the most fluent round of jumping. Dear Susie walked, stepped and occasionally jumped the obstacles depending on my confidence level but we got round – with only a couple down – and enjoyed ourselves hugely. She had looked after me so well.

We entered the next show, a couple of weeks later. This time we were a little more fluent and had just one fence down! Imagine my amazement when I was handed a yellow rosette later in the evening. I hadn't even considered that we might be in the ribbons but apparently there were only two clear rounds and almost everyone else had been eliminated for stopping. There were one or two interesting jumps, freshly painted in anticipation of the Champions of Scotland to be held at the yard in late July. Susie hadn't noticed them but others clearly had!! My first jumping rosette – just two weeks after my 50th birthday!! What a pity we missed the prize giving.

Inevitably as she established her place in the pecking order out in the field, Susie got kicked, fortunately not seriously, on a hind heel. It was awkward, though, to keep it clean and we certainly didn't want flies near it so she was confined to barracks for a few days. She was sound so I still rode her every day out in the field on the clean, lush grass with a bandage protecting the cut. Like Chelsea, she is a gem to nurse and never complains at our ministrations. Whilst confined to her box, she was happy to watch the stable girls as they mucked out and went about their business. They, in return, kept her amused with hay, treats and lots of pats as they came and went by her box. She quickly became a yard favourite.

In late July, the Champions of Scotland took place and half our riding field was transformed into a lorry-park and site for temporary stabling whilst the show field itself became two jumping arenas and a mini tented village. What would Susie make of this, I wondered? The answer was – very little. She was quite happy to be ridden round the lorry park, completely unbothered by children playing football, some-one riding a motorised scooter and the general hubbub of a busy show. We still had plenty of space to ride so our routine was undisturbed as she certainly didn't mind sharing the riding area with the competition horses. One thing amused me, though. She definitely thought she might be jumping and was quick to tell me that she thought we should be warming up in canter! Like Chelsea, she clearly enjoys shows although I suspect in Susie's case, she regards them more as social events than competition!

Lessons with Sue were going well. For a while we were able to work outside, but as the days shortened we moved inside. October brought my biggest riding challenge to date – jumping indoors. Unlike most riders, I have never been very confident in the confined space of an indoor arena. It took a long time before I was brave enough to canter inside and the prospect of jumping filled me with terror. I was fortunate to get a place on a clinic being run by Andrew Hamilton on one of his visits to Loanhead. Andrew is one of Scotland's leading show jumpers and is one of these rare people who can both do and teach. I had already attended one of his clinics outdoors with Susie and really liked his confident and encouraging manner.

With legs like jelly I entered the indoor school at the appointed hour, filled with dread and even more terrified to find a full course of jumps set up. Andrew asked me to pop across a tiny cross pole. We were simply dreadful. Or at least I was! Dear Susie clambered over as best as I would allow her. The other two in the class jumped effortlessly round the course. Then it was my turn. My brain was scrambled and I told Andrew that I was completely overwhelmed. He suggested that I go for fences one and two then see if I could carry on from there. "After all," he said, "If you can jump one fence, you can string a few more together!" Mercifully Susie knew what she was doing and set off round the jumps at a steady canter. We made it over the first two and then headed off for the next... and the next... and so on round the course. A clear round! We finished to rapturous applause. The audience knew of my fears and appreciated the achievement. I retreated to a corner of the school to watch the others, still shaking like a leaf. Once they were finished, we went round again and this time, I have to confess, I felt much happier. Another clear round and my demon was banished.

During October and November, Susie and I continued to jump indoors and my confidence grew. Dear Susie knows that her job is to stay under me at all times! So often she has helped me out and her cleverness never ceases to amaze me. On one occasion we were coming into a grid with three fences and the instructor suggested

I make some adjustments to my position as I approached the jump. I didn't quite pull it off and came into the first fence rather off balance. I wasn't quite sure what would happen at the second element. Would she stop or would I fall off? Neither happened. She simply adjusted to scoop me back on board, fiddled with her stride to take the jump and then adjusted again over the last element. We were through clear and I was still on board. We had enough time to get back together to reach the next fence on a good stride. What a kind and clever horse and worth her weight in gold in a competition, for although our style was not the best, the treble stayed up so no faults would have been incurred.

At the beginning of December, we decided we were ready for our first competition indoors. I dug out my show jacket and boots. I was quite used to riding in my boots but hadn't had a chance to practice in a show jacket as we hadn't needed to dress correctly for the outdoor summer shows. We gave ourselves plenty of opportunity to warm up. I soon forgot about wearing a jacket as Susie effortlessly soared over the practice jumps. We had opted to go early so we were soon called in to start. We flew round, or so I felt, and went clear. We would probably have a jump off to think about. Indeed we did although there was just one other clear round, a riding club friend on her very nice young horse. We were certainly in the ribbons. We were drawn first. It was the full course again and my intention was to take the same approach as last time. I vaguely remember someone saying "No faster than the last time" as we cantered round to the start. I may have heard but Susie didn't! She knew the course and off we went. Clear again and a good time. So good that I nearly fell off as we stopped after the finish! It was enough to win.

As I rode in to take our place at the head of the line I had a huge smile on my face and a big lump in my throat. Our first win with a double clear at that! Susie loved the prize giving and was delighted to have the lovely red rosette on her bridle! As the show jumping winner's theme tune started and I was invited to lead off on the lap of honour, my eyes filled with tears as my childhood dream came true. Fortunately Susie knew exactly what to do! Aidan was proud as punch too and we all had a victory hug outside.

When I first bought Susie, I wondered what Chelsea's reaction might be as she was no longer an 'only' horse. I needn't have worried. She understood from the start that Susie was part of the family and there was enough love to go round. She does, however, cherish her time out in the woods – her special time with me alone.

Chelsea's idea that she is Queen of the Gallops was made clear to me again one glorious autumn afternoon. I came to the yard with the intention of schooling Susie but Chelsea had other ideas. She nickered to me when I arrived and came over to stand with her head over the door despite having a full hay net. I started to get Susie ready, but Chelsea's eyes never left me. "Come on", she said. "We could have such fun out there." Partly out of sheer curiosity, I fetched her tack. She smiled broadly and nosed her way into her bridle. She really did want to go out. I actually felt very humble and hugely honoured. She had been out in the field all day and now she had a full manger of hay but what she really wanted was to take me out for a gallop. Who says horses don't enjoy being ridden?

It was a wonderful afternoon. There was a nip in the air and a good cut in the ground after the hard summer going. Chelsea's coat was gleaming copper in the sun. It made me realise how truly she is a hunter, loving the cooler days and the give in the ground. She was right. It was just glorious to be out on a beautiful afternoon with a wonderful, willing companion. I told her how super she was and how much I loved her and her ears flickered back to me, asking me to keep telling her these wonderful things, even if she did know them all already! As for Susie, she didn't mind at all. She had plenty of hay and had enjoyed a brush and a cuddle and in any case, she knew well enough that there were bound to be carrots for her too when Chelsea came back.

Chapter 10: *New Directions*

April 2002 brought my work in London to a close. As ever, I was sad to leave the many friends I had made. All my projects were in good shape and I was touched at the many thank you notes I received. At last, however, I had the opportunity to set off in my new direction. I was excited about this – although it was also quite scary. Hitherto, my work had always been in the form of large contracts, generally extending over at least a year. I would usually only have one or two clients at any time and they always found me through word of mouth so I had never really needed much in the way of sales and marketing expertise and our cashflow was very predictable.

Now things would be different. I needed to work on my selling and marketing skills as we would need to build a larger customer base for our new work of coaching and facilitation. I was also keen to work in and around Aberdeen, at least for a while. I'd had twelve years of regular travelling and wanted a break from it. I would need to build up my contacts and do more networking as, up until now, I had done very little work in the local area. So much to think about...

I also had an immediate project from Aidan. In January, he had finished the text of "Gentleness" and now we had to get it published. Whilst in London, I had picked up a newspaper on a train and, to my amazement, found in it an article about Publishing on Demand. This was just what we needed. I contacted the company mentioned and was delighted by their friendly, helpful response. Our next step was to organise the cover for the book. It felt such a small thing after the effort of writing the book but in fact it took us almost as long!

Again, the power of synchronicity brought us into contact with a wonderful equestrian artist who was delighted to help us with the cover. She did some sketches based on photos of Chelsea and then came to visit both horses. It was not long before we had the perfect book cover. In the end it was so simple. Fiona produced some wonderful, quite large pastel drawings and, just for fun, I photographed them. When I looked at the result, I realised that it was just what we needed. I played about with it adding the title and Aidan's name to the photo and there was our cover. At last we had everything we needed for the publisher. The galley copies started to arrive and we proof read them diligently until we were satisfied everything was just right. Then, in time for Christmas, the author's copy arrived. Holding "Gentleness" was a very special moment.

It was a great end to the year as we really felt we had begun to make the shift we wanted. Another highlight had been the Secretan Gathering in October, now an annual event for us. This year we had offered to help. Our biggest challenge was providing decorative panels for the main meeting room. The Gathering theme was the Castle – which Lance Secretan used as an acronym for Courage, Authenticity, Service, Truth telling, Love and Efficiency, being his key attributes for leadership. Helped by Ruth Webber, a dear friend who is also a talented artist, we designed and painted 8 king size bed sheets which, when draped around the room, would look like the walled garden of a castle.

It was another great piece of learning for me. I had always felt that I couldn't draw and dreaded 'creative moments' in workshops. With Ruth's gentle help, I found that actually I could be quite artistic and, more to the point, really enjoy the experience. Ruth and I did most of the planning, initial sketches and broad layout of the final sheets. Aidan joined in and produced many of the final panels, adding detail to our outlines. The sheets were painted in our garage which was the only space that was big enough and could cope with the occasional spills and splashes of paint.

Our other contribution to the Gathering was a session entitled "Gentle Leadership". This was our first venture into bringing the horses into our work. Presenting this at the Gathering was perfect for we knew we were among friends who would be supportive and yet could be trusted to give us honest feedback. Ours was one of the breakout sessions and we were intrigued to see who would choose to attend. Preparing the session was pure joy. We used PowerPoint slides with photographs to illustrate the main learning points and then had some exercises so that everyone had the opportunity to participate. We wanted to bring the spirit of horses into the room and so I made six large collages of horse pictures, each with a poem at the centre. Already my new found creativity was coming into play and I also had a very good reason for buying every horse magazine I could find for about three months!

During the early morning meditation before our presentation, I was aware that Chelsea and Susie were with me and as I started our session, I felt them with me again, one on either side. It was as if I was speaking for them as well as myself. Our allocated ninety minutes flew by and we only managed to deliver about half of what we had prepared to a very packed room. It didn't matter. The feedback was amazing and we were deeply touched as well as being very encouraged for the future. I also became aware that my purpose in this life is "to speak for horses".

Over our Christmas break, we had our customary look back over the year. Not only had "Gentleness" become a reality but we had also delivered our first "Gentle Leadership" workshop and most of our work was coaching and facilitation. The shift we dreamed of on the Secretan programme two years before was really beginning to happen and we finished the year confident that our new venture would be a success.

The only cloud on the horizon was Sue's departure to Australia. She had visited the country on a number of occasions and felt that she could do well there. We bade her farewell with a heavy heart. During the time I had been having lessons with her, she had become increasingly interested in my work and its application to riding. She enjoyed taking a 'coaching approach' to our lessons and had started to use it with other pupils too. I knew she would be hard to replace but I wished her well as she set off to pursue her own dream.

Chapter 11: *Gentle Leadership*

As always, we were looking forward our annual break in Fife. Early January was usually a quiet time work wise as people settled back into their daily routine after the Christmas break. We enjoyed the opportunity to relax and plan for the coming year. As usual we went via the yard to say goodbye to the horses. When we arrived they were out in the field. We went to bring them in and were horrified to discover that Susie's left eye was closed up and had swollen to the size of a cricket ball. We called the vet at once and he arrived quickly. He was clearly concerned and administered a cocktail of painkillers, anti-inflammatory drugs and antibiotics. We'll never know what really happened but it seems that she had banged her head in her stable overnight and injured her eye. Knowing that we were about to go on holiday, the vet and the yard staff reassured us that they would administer the necessary drugs and stay in touch with us by phone. We were only going to be about 90 miles away so I decided to come back during the week to meet the vet and check that Susie was responding to treatment.

We enjoyed our break, relieved to know that Susie was responding well to the medication. As well as planning our year, we began to think about how we would launch "Gentleness". A number of our North American friends were coming over in April for a conference that we too would be attending so we decided to hold the launch the weekend before. We found a venue – a lovely old house on the edge of Ballater which was being converted into a small hotel. We would be their first event. They had an excellent caterer who prepared a wonderful lunch, complete with celebration cake, for our forty guests. We ordered copies of the book, sent out invitations and planned the 'agenda'. Our house would be full too. I drove to Edinburgh the day before to collect the Canadian contingent who were staying with us, as were dear friends whom we had known since Zambia days who drove up from Hampshire. Over dinner, we discussed the logistics for the following day with happy anticipation. The launch was pure joy and celebration with the last guests finally drifting away well after 5 o'clock. It was truly a day to remember.

"Gentleness" tells a story about a person and a horse who are both crushed by established views and the compulsion to conform – typical 'old story' leadership. Eventually they are able to experience love, inspiration and the removal of fear such that their spirits are once again free to achieve great things. Aidan used some of our experience with Chelsea for the horse thread of the book and, in the end, kept her name in the book even although the book is a work of fiction, albeit with a message.

As people began to read and enjoy the book, I began to get requests from some of my coaching clients to come and meet the horses, especially Chelsea. I love watching people, many of whom had never had any contact with horses, connect with Susie and Chelsea. Susie is such a sweet creature that even those who were a little afraid of horses were soon happy to be stroking her. I am amazed at their observations too – the lie of her coat, the softness of her mane and tail. These were things I had taken for granted but now I too saw them with new eyes. Mostly, though, I love to see people who arrive tired and care worn leave refreshed with a spring in their step after an hour with Susie. Coaching conversations flow easily in her stable, with the gentle mare herself watching and listening with deep interest.

Chelsea is quite different. The hard times in her past make her more wary and she is quick to appraise people. She is an excellent judge of character and very quick to spot those who are not congruent – who are saying one thing and feeling something else. The interesting thing about her is that people want to be liked by her. Most people sense they are being appraised as she looks at them with her deep, wise eyes and they desperately want to pass the 'Chelsea test'. This too is a great opportunity for learning. When asked what might improve a situation, so many people will say "If only he would do this... or she would behave like that... things would be better." Yet, here they are prepared to do almost anything to win Chelsea's favour. What might they change about themselves that could help their situation at once? After all, if they are willing to change for a horse... It was Ghandi who said "Be the change you want to see in others" and Chelsea has inspired more than one person to do just that.

In July, Sue arrived back from Australia for a month's holiday. I had been gradually rehabilitating Susie after her eye injury, which sadly resulted in partial loss of sight in her left eye. When I started to work with Susie again in February, she was reluctant to do anything other than walk. Very gradually we moved up through the paces as she adjusted to what she was seeing. Unlike Chelsea, who is so tough and determined to be well, Susie needed more patience and gentle encouragement. By the time Sue came home, Susie was happily trotting over poles on the ground. Now we would find out if she could jump again. I was so glad to have Sue there on the ground for I knew she was the best person to set things up for success. We trotted over some poles to warm up and then Sue placed the poles to fit Susie's canter stride. Susie bounded over them, albeit a little reluctantly at first. Then, a wonderful moment – as we turned to come up the line of poles for the fourth or fifth time, her ears pricked and I felt her carry me confidently towards the poles, just as she used to do. She was ready to jump. Sue turned the last pole into a little fence and Susie cleared it with all her old ease and grace. Her smile was even bigger than ours as we all had a hug.

While Sue was here, I went on a weekend course in Equine Assisted Psychotherapy which was taking place in Edinburgh in August. Although I had been reading all I could, meeting up having some more formal 'live' learning was just what I needed and although it was focused on therapy, I thought it would at least give me some ideas for developing an Equine Assisted Learning (EAL) programme. It was an interesting experience – me and these eleven therapists! Being analysed by them constantly was a little overwhelming but the course itself was very helpful and I saw at once how I could develop a programme for learning. It inspired me all the more to do this work. The model we were taught was based on having a therapist and horse person work together. I could see the value in this from both a learning and a safety point of view. For us it would be a coach/ facilitator and horse person but otherwise the model and many of the activities worked perfectly for us in our learning context.

Sue was eager to hear all about it when I got back. She had been having second thoughts about whether Australia was really for her. The prospect of doing more equestrian sports coaching and working together to do EAL appealed to her. Chelsea, Susie and I demonstrated the activities to her and showed her how it could work. She agreed to join our team. She went back to Australia to start arranging to come home to Scotland. I was delighted. I would have my teacher back and I had also found the perfect partner for my EAL work.

Once Sue came back in October, she needed some time to settle back into life in Scotland but that didn't stop us talking about what our programme might look like. I also turned my thoughts to a name for our new service. We had always used 'gentle leadership' to describe the work with the horses and thought that we could use that. We needed to devise a logo so we approached a local graphic designer. Lynn had read "Gentleness" and loved horses so we knew she would come up with just the right design. It was she who said, quite out of the blue, "Why don't you just call your company **gentle leadership**? It's a perfect description of who you are and what you do." It was an inspired idea. We had been feeling for some time that we had outgrown 'Midpark Consultants', our original name which had served us so well initially. These days, however, we weren't consultants any more. Lynn designed a beautiful logo and we did the necessary administration to change our company name although we still needed a name for our EAL programme ... but more of that later.

At last everything was coming together. When I had originally bought that sad little chestnut mare, I never imagined how she would change my life. But I think she had it planned all along.

Part II: *Perspiration*

Chapter 12: *Getting Started*

After finishing in London, I had taken some time out to think about "What next?" Although I had enjoyed working on large projects that typically lasted for several years, they generally meant a lot of time away from home. They were also very intensive and, although satisfying, I was beginning to find them quite draining. Perhaps I was just getting too old for them! Clearly it was time for a change. I wanted to spend more time at home with the horses and, of course, to bring them into my work.

When I reflected on what I had most enjoyed about all of my work to date, it was undoubtedly watching people and organisations grow and develop. I have been a coach for a long time – since well before it became fashionable – and so I decided to concentrate on offering coaching and facilitation to businesses and individuals with the work with the horses as an 'extra' for clients who wanted some experiential learning.

I took some time out to brush up both my coaching and facilitation skills as I had not really kept these up to date since moving back to Scotland. I also increased my networking efforts in Aberdeen. I had always done a certain amount and had served on the Committee of the Grampian branch of the Institute of Directors for several years but I knew I would need to do more. I became involved with Business Networks International (BNI) who were starting up in Aberdeen at this time and became the founding director for one of the new chapters in the city.

As I networked, I started to talk about my plans for introducing the horses into our business. The feedback was generally along the lines of "Are you crazy?" From this very informal survey, it was clear to me from the outset that whatever we did with the horses had to be well thought out and presented in a totally professional and business-like manner. I also realised that, as learning with horses was a completely new concept here in Scotland, I would also have to make a market for it.

Thomas Edison was right when he said "Success is 10% inspiration and 90% perspiration"! I'd had my inspiration. It would need some perspiration from now on if I was going to make my dream a reality.

As a business person, doing all the planning, marketing and other related tasks made sense and yet I was also uneasy. I knew from my own experience with the horses and from seeing how they changed the people who came to visit them that they offered much more than just effective learning. Their magnificent presence and spirit seemed to bring out the best in people and provided the perfect environment for some very significant conversations. By simply talking about the business benefits, I felt I was selling them short. In the end, however, I decided that if I wanted to make learning with horses credible, I would have to concentrate on defining the service and benefits in business terms and trust that participants would get the rest anyway.

At least I was on home ground for this task, having helped many businesses develop both strategic and operational plans. From the outset, I have always had a plan for our business and I also knew the value of writing it down. I always believed the process of writing made a difference but it came home to me very clearly when I was clearing through some old papers and came across the very first plan I did for Midpark Consultants when we set up the business in 1990. Our five year vision was:

- To move back to Scotland
- To continue to look after our customers in the South (of England)
- To build up the business in Scotland

That was exactly what we had done when we moved back to Scotland in 1994.

I started at once on our business plan for **gentle leadership**, adding in the activities needed to bring the horses into our work. Although it was really only planned as an 'extra', there was actually a lot to do if it was going to have any chance of success and at times it all felt quite daunting. Happily, there were always people willing to help.

Although BNI didn't bring us a great deal of business, it brought us into contact with several suppliers who are still with us and who were integral to getting **the red horse speaks** off to a good start.

Lynn Clarke of the Design Room is our graphic designer and was a fellow chapter member. Her first piece of work, mentioned previously, was to design the logo for **gentle leadership** and to inspire us to change our company name.

Kenneth McKay of Initiative 2 belonged to another BNI chapter in town and was also a member of Aberdeen Entrepreneurs. He had ridden for many years although he had given this up as his business grew. He was genuinely interested in our new venture and so he became our web designer.

It was through a visitor to our chapter that I was introduced to Liz Marchant of Marchant Communications who initially did all our Public Relations (PR) work. I had been told that Liz did some training and development and it was arranged that we should meet for a coffee. As it turned out, Liz didn't do training and development but was interested in communication and expanding her PR business. I told her of my plans for **the red horse speaks** and, like Lynn and Kenneth, she 'got' what I was trying to do and was very keen to help. She became our PR person.

BNI also gave us our first PR opportunity. We decided to make our last chapter meeting before Christmas a family event and invite spouses, partners and other family members who had helped us make the early start to attend our meetings each Wednesday morning. I mentioned to my fellow members that I would bring one of my horses and of course they thought I was joking.

Susie's visit to BNI was a great success and rated a photo and a few words in our local newspaper. It also taught me a lot about what the media are looking for and how to work with them successfully. Sue and I rose at 5.00am in order to feed Susie and plait her mane with tinsel before taking her down to the hotel where our meetings were held. Whilst it would have been perfect to take Chelsea as our 'red horse' we knew she would find the occasion stressful as she would always have been wondering when the competition was going to start! Susie, on the other hand, regards outings as primarily social occasions where taking part and meeting her friends are much more important than winning. An hour in the hotel car park was not going to cause her any stress at all. The hotel manager had reserved us a parking space for the horsebox in the middle of the car park opposite the front door. It was perfect for Susie and she had a great time being patted and fussed by everyone who came in and out of the hotel. She was also a welcome surprise for the local children who walked past the hotel on their way to school. As we kept a discreet but careful eye on her, we heard several conversations along the lines of:

"Mummy, mummy, there's a horse in the car park!"

"Don't be silly."

"MUMMY, THERE IS. THERE IS."

And so they would come into the car park and give Susie a pat.

I looked tidy enough in trousers, polo shirt and my **gentle leadership** fleece as we had agreed a relaxed dress code for this meeting. I hadn't really given much thought to the interview with the reporter from the local paper which Liz had arranged. After the meeting, however, we met the photo-journalist back at the yard. He wanted a business slant on our story and my casual attire was not suitable for the photograph he had in mind. Fortunately, I had asked Liz to be there as I was a little apprehensive about meeting the press for the first time in a while. She loves clothes, is very glamorous and is always wonderfully well dressed. That morning she was wearing a very smart winter coat with a large fake fur collar – and even more fortunately the coat fitted well enough for me to wear it for the photo shoot. I was beginning to see how the mix of business and horses could be taken into all aspects of my planning.

Although there were still those who thought my new venture was crazy, I had a strong sense that this was what I was meant to be doing – and so I kept going and started to do the detailed planning that would bring **the red horse speaks** to life.

Chapter 13: *Defining the Service*

One of the first things Sue and I needed to do was to define this new service and decide how we would work together to deliver it. To do this in a meaningful way we needed to have Sue experience it for herself, especially to understand the role of the horse specialist in our partnership. Although I could explain it to her in some detail after my course, I knew it wasn't going to be enough for her. I had already discovered that although my facilitation skills were very good, I needed to make some changes for them to be equally effective in the context of working with the horses and so I knew it would be the same for Sue in respect of her horse skills.

Fortunately, we found a conference about therapy with horses which was to be held in March in Nashville, USA. This worked perfectly for me as I had also planned to go on a personal leadership retreat in Austin, Texas which started a few days after the conference finished. Happily Sue didn't mind a trip to America. Although the conference concentrated almost exclusively on therapy with horses, I knew from my course that we could easily translate the activities and context from therapy to learning.

The conference was great fun. We spent a whole day at the local County Show Ground, which had a huge, heated indoor arena where we watched several demonstrations that covered all the practical aspects of this work. Sue quickly grasped what her new role would entail – and recognised that working with the horses in this way would be quite different from her usual role of equestrian sports coach. Happily she was up for the challenge and the 'unlearning' that would be required. The rest of the time we went to classroom sessions and learned just as much from listening to others' experiences. The whole of the final morning was given over to a lengthy lecture on Schizophrenia followed by a discussion session which Sue and I decided we could easily miss. We wanted to spend the time doing some planning while everything was fresh in our minds.

We went off in search of a quiet spot where we could have a coffee. As we crossed the hotel lobby, we noticed two other 'truants' from the lecture and recognised them as the people who had given the practical demonstration we liked best. We smiled at each other and they invited us to join them and soon we were happily chatting. It was one of these wonderful occasions when we quickly felt as if we had all known one another for a long time. Since then, Ann Romberg and Lynn Baskfield have become good friends as well as fellow professionals in the world of learning with horses.

Our coffee with them was hugely encouraging. They were about a year ahead of us and were very generous in sharing their experience to date. Ann's background was very similar to mine in that we had both spent some years working for large organisations before setting up our own businesses while Lynn and Sue shared a common background of having been involved with horses since childhood. We came away feeling very much more confident thanks to their generosity and kindness.

One of things I am convinced about is that when we have a clear sense of purpose and passion, there is no such thing as a purely 'chance' meeting. The Universe is quick to bring together those whom we are meant to meet. This was so beautifully expressed by W.N Murray in 'The Scottish Himalayan Expedition':

"Until one is committed there is hesitancy

The chance to draw back, always ineffectiveness.

Concerning all acts of initiative and creation

There is one elementary truth, the ignorance of which

Kills countless ideas and splendid plans:

The moment one definitely commits oneself,

Then Providence moves too.

All sorts of things that would otherwise

Never have occurred. A whole stream of events issue from the decision

Raising in one's favour all manner of unforeseen incidents and

Meeting and material assistance,

Which no man could have dreamed would come his way"

I found this quotation in Joseph Jaworski's book "Synchronicity – the inner path of leadership" which has long been one of my favourite books on this topic. The quotation too is a firm favourite which I find myself re-reading in moments of both gratitude and doubt.

The conference and our conversation with Ann and Lynn enabled us to determine the broad framework of our new service. Our target market would be corporate clients, at least initially. It was a market I already understood and to which I had access through the rest of my work. We were also very clear that what we were offering was learning and not any form of therapy. We would concentrate on leadership and team development. We knew that Aberdeen and north-east Scotland was a fairly conservative area so it could take some time to build our market there and even when we did it was still a relatively small area so we decided that our model had to be portable, allowing us to work in other part of Scotland, UK and indeed the world. That meant that we could not rely on specially trained horses and would have to set some standards for selecting venues. That was not too difficult for us as Sue has a great network of contacts from all her years as a competitor and now as a coach. Being able to work with the horses we are given is also important to Sue and me in terms of walking our talk. Many of the teams we work with are assembled for their skills rather than for their personalities and yet the team members need to be able to get along together. By showing that we can work with whatever horses are available allows us to explore relationships in a very practical way as part of our programme.

Horses, of course, are central to this work and so we also spent time thinking about how we wanted it to be for them. At the various demonstrations, we had seen horses being subjected to treatment that left both of us feeling very uncomfortable. During activities, they had been cornered and had heavy saddles thrown on their backs. Rope halters were also much in evidence. These halters have pressure knots and so need to be used with very sensitive hands by people who know when to tug and when to release otherwise the knots can hurt the horses. Some of the sessions were very thinly disguised horsemanship lessons, usually based on the work of either Monty Roberts or Pat Parelli. Whilst we have high regard for the methods of these well known horse trainers, we don't feel their methods are safe or appropriate for equine assisted learning as they are actually designed to train horses rather than people and whilst their equipment may look very smart and apparently very simple, it is actually quite specialist and designed to be used by people who have had at least some training.

All of this led Sue and I to decide that whatever equipment we used during our sessions had to 'safe in unskilled hands ™'. For example, all our own horses and those with whom we work elsewhere wear traditional English headcollars which are broad and soft so that if they do get tugged by clients, any pressure is well distributed and doesn't hurt the horses. We very rarely allow clients to put on and take off the headcollars as even the most patient of horses is likely to become head shy with rough handling, however unintentional it may be. The horses wear head collars throughout the workshop and clients are requested not to hold or pull on them. When we long line or lunge it is always from a well fitting headcollar. We never use a bridle.

We have some activities that involve participants putting a rug on a horse - as an exercise in giving instructions rather than to see whether they can put a horse rug on correctly - and for this we use a soft fleece rug so that even if it gets tugged about a bit, it is never uncomfortable for the horse and will not hurt its back. We simply don't do any activities where the horse is likely to get trapped in corners with poles or other similar objects. In fact, it's one of the few rules we have - and that is that the horse always has to have somewhere to go.

Both Sue and I take an individual approach to working with our respective clients and so we decided that all our workshops with the horses would be tailored to meet individual client's needs. This suited us personally as neither of us would be very good at delivering exactly the same programme over and over again. We thought it would be more interesting for the horses too. It also offered more possibilities for repeat business as we could increase the difficulty of the activities and introduce different horses.

Although our longer term aim was to be able to work anywhere, we decided that initially we would work with Chelsea and Susie with groups of up to twelve participants as that felt manageable and safe for us. We chose a core of about six activities that we knew well and which could be used in a variety of contexts.

We were very fortunate that the owners and staff at Loanhead, the wonderful livery yard where I kept Chelsea and Susie, were hugely supportive of our new venture and were happy to host our events. In fact it has worked well for them as we usually use the indoor arena during the daytime when it is quiet anyway, as most of the people who keep their horses at Loanhead are either at work or at school during the day. Thus it has provided them with some additional income.

Alongside all of this, I was speaking to Liz about our plans and how to develop our PR strategy. She suggested having a formal launch event and that seemed a great way to get started and gave us a focus for our efforts. We decided on a date in September so that we could get invitations out before the summer holidays and hopefully have some time after the launch to do some workshops before winter set in. Subsequently we have found that our workshops can run most of the year. There is a natural break in December when everyone starts to think of Christmas and we tend not to plan much for January as the days are still so short that we could be starting and ending in the dark!

The only thing that I hadn't yet worked out was what to call our new service now that **gentle leadership** had become our company name. Telling people that we now offered 'equine assisted learning' was not going to be either meaningful or memorable so this new service needed an identity of its own.

The solution came unexpectedly one Saturday morning. After the Nashville conference I had spent a week at a personal leadership retreat in Austin, Texas. It was an opportunity to think more about the life I really wanted. The retreat placed special emphasis on what was described as 'my 100 year legacy.' I came away with even more certainty about my work with the horses and greater clarity about overall direction. As is often the case with such events, new friendships were forged and I still keep in touch with some of the people I met there. After the retreat one of the women, named Reggie, with whom I had some interesting discussions about coaching, had gone to stay with friends in another part of Texas before returning home to Boston. When she got back, she emailed me to tell me about her visit and mentioned that she had brought a small painting of a 'red horse, standing in a garden'. (a red horse in America is what we in UK would describe as chestnut) She described it as being rather whimsical but it had caught her eye and so she decided to buy it.

As mentioned previously, one of my personal sources of inspiration is the Native American animal cards. I first came upon them at one of the Secretan events and have worked with them ever since as a focus for meditation. Native Americans believe that animals – and indeed all living things – bring messages and learning to humans. As I replied to Reggie, I found myself writing "You might be interested to know that the red horse speaks of..." and immediately I knew what to call our new service. Of course, **the red horse speaks**. It fitted perfectly as "the red horse speaks of learning being more effective when it's fun and the need to balance work and play." It was perfect. And, of course, it also honoured my own red horse, the wonderful Chelsea, who had spoken so clearly to me and would do so to many others. I finished the email to Reggie and thanked her for the inspiration.

Over the spring and summer of 2004 as I continued with my weekly lessons with Sue and Susie, we found ourselves talking more and more about **the red horse speaks**. I started to tell Sue more about my work so that she could start to understand more about the business world as she was quite apprehensive about working with business people, feeling that she didn't have much to offer them. We also practiced our coaching skills and came up with some great ideas to incorporate into our programme. That time spent together helped us to develop our partnership as we worked together to continue to help Susie adapt to her partial sightedness.

There were also some practical things to do. We sorted out insurance, which was an interesting exercise. We knew that it was important to distinguish **the red horse speaks** from any form of riding school or riding instruction and to be clear that all our work was on the ground. We were incredibly fortunate that our insurance broker, a delightful Irish woman, also had a good knowledge of the horse world. She very quickly understood what we were offering and soon found us a good deal which saw our professional indemnity insurance increase only very slightly.

In addition to my professional indemnity insurance, Sue also had her own cover which was well in excess of anything we might need as it enables her to teach both eventing and dressage to a high level as well as working with Pony Club and more 'ordinary' clients such as me. One of the items she has subsequently added is a 'care and custody clause' which covers her when we are working with other people's horses as happens when we are delivering workshops away from home.

Another consideration was First Aid knowledge. Sue had had some first aid training but completed a full course run by St John's Ambulance Association. We thought this would give our clients confidence that we were well organised and responsible. Later on I also completed the British Horse Society's equine specific first aid course which concentrates on treating people injured as a result of equestrian activities. Although it focuses mainly on the consequences of riding accidents, I also found it to be a good all around first aid course.

I have been passionate about the inclusion of people with disabilities into everyday activities ever since my experiences during 1980 which was declared the International Year for Disabled People under the patronage of Lord Snowdon. At that time, I was working for a Social Services department and volunteered to organise some events to raise awareness amongst my fellow administrative staff. These events had a profound effect on me as I realised how often a very small effort made a huge difference and ever since I have done everything I could to ensure that any event in which I am involved is accessible to people with disabilities. Naturally, therefore, I wanted to be sure that whatever we did with **the red horse speaks**, it would be possible for disabled people to join in. We were very fortunate that Sue knew the leader of a local Riding for the Disabled group. They were interested in what we were doing and decided it would be a good 'off-season' activity for their helpers. We mentioned that we would also value their advice on how to make the activities suitable for disabled people and so they invited some of their clients to come along as well. It was hugely helpful and we soon discovered how to modify our activities for people with limited mobility or those who were wheel-chair bound. We also learned about which surfaces were best for wheelchairs.

As **the red horse speaks** is a form of experiential learning, I spent some time finding out more about this form of learning as well as its benefits. I decided to stick to a very simple but well tried and tested model for our work, namely Do, Review, Learn. It works like this:

1. **Do.** We set the activity for the group and as far as possible leave them to plan and carry out the activity by themselves. We might ask questions but will only make suggestions as a last resort.

2. **Review:** After each exercise, we review the activity. Sometimes, clients want to try again, using a different approach and we encourage this. They may even want to

experiment with a variation on the exercise and we encourage this too – provided it is safe to do so.

3. **Learn:** This final stage is the link back to work or life. How will the learning be applied? This is captured via flipcharts on the wall or through the use of workbooks.

I also wanted to bring in the solution-focused approach which I use successfully in my other work and which seemed particularly apt in an environment where we wanted people to have fun and be creative. We set out 'the rules' at the start of each workshop. The first two are also great prompts during activities.

1. If it's working do more of it

2. If it's not working, do something different

3. There's not one right way, there are many

4. There are no mistakes, only learning

This approach also makes the workshops comfortable for the non horse people as it allows us to emphasise that the event is not about teaching horsemanship. We are not looking for the 'right' way to do things – e.g. to lead a horse. Conversely, this can be a challenge for the horse people as they are often constrained by their knowledge and sometimes find it hard to think outside the box.

I also did some more reading about the benefits of working with horses and decided to focus on the three main characteristics which I felt would resonate with business needs and were easy to explain:

1. Horses are individuals with their own personalities. They can be difficult, stubborn, playful or helpful. What works with one may not work with another. In other words – they are just like our colleagues and family members. The horses always seem to inspire people to try out different behaviour which they can then take back to the workplace.

2. Horses mirror back to us what our human body language is telling them – an important lesson since about 80% of communication between humans is non-verbal – i.e. our body language. Humans may not always be very good at noticing this in themselves and others but horses are acutely perceptive. They are also very quick to notice the power of our intention.

3. Horses are totally honest and have no hidden agenda and so we can trust their feedback. People take (and act upon) feedback from horses that they would not take from another human. That the horses are living beings and are therefore able to give feedback is one of the main distinguishing features of this form of experiential learning as compared to high ropes and other forms which are based on equipment.

Now that we had a clear idea about what we were going to do, how we were going to do it, had a name for the programme and a launch date, we could move ahead with the next level of planning.

Chapter 14: *The Launch*

We chose Thursday 2nd September for our launch. As far as we knew, it didn't clash with events run by any of the networking groups or with any other significant events in town. So now it was all systems go!

Having a date focused the mind as we worked out what would need to be in place by then. First of all, of course, we had to book Loanhead, catering and a marquee for our chosen date. We decided that the format for the evening would be a 6.00pm start (so that people could come straight from work) with coffee served on arrival followed by a demonstration of **the red horse speaks** in the arena at 6.30pm and then back to the marquee for wine and canapés, finishing up at about 9.00pm

My project management skills were put to good use. We wanted to issue the invitations before the summer holidays which, for us, meant they had to be out by the end of June – so we had to have our logo, invitations and the website all done by then. The website was important as it would give those receiving an invitation more information about what they might expect at the launch and, of course, would be there for anyone else who was interested or just curious. We also decided to have a working wardrobe of branded clothing to maintain our business image and so we chose polo shirts and sweat shirts complete with logo to add to the waistcoats which Sue and I had already purchased.

The website needed photographs, of course, and ideally some testimonials so Sue and I decided we should have a couple of practice sessions at Loanhead that would allow us to test out our partnership, give the horses a taste of the real thing and provide the necessary material for the website.

We decided that our key suppliers – Lynn, Kenneth and Liz - would be good guinea pigs as it would hopefully help them understand exactly what we were doing. They were all very enthusiastic. Kenneth asked if he could bring along Dave, his technical specialist who would actually be designing and building our website. We also invited Chris Oliver, a friend and client who had recently set up an exciting new business himself, the health and safety representative from our local British Horse Society branch to give us any guidance he thought necessary, and Ruth Webber, our artistic friend who had been one of the very first people on whom I'd tried out some of my original ideas. In addition, Lynn brought along her six year old son Matthew and his friend as she hadn't been able to find a babysitter – so we even had the opportunity to work with children!

At the first trial session, we started off by explaining to our guinea pigs briefly what **the red horse speaks** was all about and why it was effective. We had seated our guests in the viewing gallery and Sue and I stood down in the arena with the horses. As I started to speak, Chelsea immediately came up to stand between Sue and me, making it very clear that she was definitely a part of all of this. That evening was a dream come true for me as I prepared to work with my beloved horses for the first time and I was finding it quite hard to speak anyway. As Chelsea came to stand with us, my eyes filled with tears. Fortunately I was amongst friends who knew how much this meant to me. I gathered myself together and we started our planned programme for the evening.

At first we left the two children in the viewing gallery so that they would be safe but we couldn't help but hear some of their comments about what the adults were up to. At one point, the adults were finding it difficult to get the horses to move. It was so obvious to the children and so we invited them to come and show us. These two small boys simply marched up to the horses with such determination that the horses immediately moved exactly as requested! Thereafter the children joined in for several of the activities. As we had hoped, there was laughter and also deep learning. Sue had brought her horsebox and the living area served as our 'reception suite' where, over a glass of wine, we and our guinea pigs talked over the events of the evening. The feedback was excellent, the health and safety man was happy, the horses had loved it and we only had a few wrinkles to sort out before doing the same again a couple of evenings later for another practice session.

As we had hoped, these sessions were not just great practice for Sue and me but also gave Lynn, Kenneth and Liz a really good grasp of what we were doing. Lynn produced beautiful artwork and stationery for us, Kenneth and Dave came up with a stunning website design which soon became a live website and Liz started to drop hints to all her media colleagues that there would be something special and unique happening in Aberdeen in September.

The summer passed quickly as we gradually made sure everything was in place. August 24th 2004 was our 30th wedding anniversary but any plans for a special celebration were put aside as we worked our way through lists of people attending, outstanding tasks and practicing our responses to possible questions from guests and the media.

The day of the launch was definitely one of best days of my life. To our delight, over 60 people had accepted our invitation – many more than we'd expected. They were a good mix of existing and prospective clients, members of our various networking groups and supportive friends. Liz had done a wonderful job getting press interest and so we hoped a few reporters might turn up for our press conference at 2.00pm.

The thinking behind having the press conference at 2.00pm was to give us plenty of time to get ourselves and the horses brushed up and looking smart beforehand and then there would be time afterwards to give the horses a break before our demo started at 6.30pm. Feeling in no need to hurry, we were sitting at home having a leisurely cup of tea before going to pack the car... then the phone rang at 9.45am with a call from Liz to say that we had a request for a photo call at noon and, by the way, a journalist from the Scotsman would be calling in 10 minutes! All of a sudden, the tempo had increased. I quickly called Sue and told her we were needed at noon.

I spoke with the man from the Scotsman who was very interested in what we were doing and then gathered my things together to go to Loanhead to meet the photographer. As I was driving along the road, a delightful journalist from our local paper, the Aberdeen Press and Journal, called on my mobile phone and so I stopped in a lay bye to talk to her. That went well too and the midday photographer was fine. It turned out that he was from a press agency and so needed to have the story early in order to sell it on to other newspapers. We managed to have ourselves and the horses looking neat and tidy for him. Fortunately Chelsea can be brushed up in five minutes but Susie, being grey, takes a bit longer. We had bathed them both the day before but Susie can usually be guaranteed to find a muddy patch somewhere! Sue had got to Loanhead before me and had done a great job.

Now for the press conference. We walked down to the indoor school a little before 2.00pm just to see if anyone was there... We were amazed. We felt like royalty with a veritable battery of cameras set up and waiting for us, including a team from Grampian TV who wanted to shoot an item for their news programme that evening.

My first thought was how the horses would cope. They had both been to small shows but hadn't experienced anything like this. I needn't have worried. They were true professionals and never put a hoof wrong. Susie was so relaxed she even had a roll, rising up from the arena floor covered in sand and dust which made everyone laugh (even Sue, who had spent ages brushing her beforehand!). We were blessed to have a small group of volunteers come along to demonstrate some activities for the press. They did a great job and were all very articulate in their responses, especially for the TV crew. I was very glad that I had spent time practicing my own responses to possible questions and thus could give a calm, confident performance in front of the cameras.

After the press left, we had arranged for a professional photographer to come and take some shots of us so that we had a stock of photos for our own use as most newspapers have copyright of the photos they take and use in their own publications. Although it meant more posing, it was well worth it as we later found that, when invited to submit articles to magazines, we would usually be expected to provide high quality photos – which, of course, we are always able to do.

One thing I did decide that afternoon – I could never be a supermodel! Although I enjoyed posing with my beautiful horses, I was not really very good at rapidly changing my outfit. Some of the photographers wanted shots of me in a suit with my briefcase and others wanted me in my working clothes. I was very glad to have the living area of Sue's horsebox as my dressing room! Happily, the horses were perfect just as they were.

We managed a short break between the photographers and the start of the launch itself. This gave us time to feed the horses, have a quick snack ourselves and check that everything in the marquee was ready to receive our guests. Lynn had prepared beautiful folders with information about our programmes and Liz had found a chocolatier in Belfast who made chocolates in the shape of a horse's head. He had made up little gift boxes containing two milk chocolate and two white chocolate heads (the nearest we could get to chestnut and grey) and tied them with a ribbon in our company colour and a label showing our website address. They made perfect 'party bags'!

By 6.00pm people were starting to arrive and we were grateful to Shona and the yard staff for being willing to serve teas and coffees while we talked to our guests – not to mention later 'guarding' all the canapés once they were set out as they had already come to the notice of the yard cat! At 6.30pm we ushered everyone into the arena so that they could take their seats for the formal proceedings. Sue brought the horses down and we were ready to go. Aidan had taken the role of Master of Ceremonies and had prepared a wonderful introductory speech then it was over to me to explain what **the red horse speaks** had to offer. I kept this brief as I knew that the horses would be able to say much more. We had chosen four activities and invited volunteers from the audience to come down and participate. By the end, everyone wanted to have a go, including an accountant friend who came straight from the office in crisp white shirt, tie and jacket – despite our suggestion on the invitations to wear casual clothes. He told me later that he was so sure he would just watch but in the end found the horses irresistible and decided to join in one of the activities anyway!

Chelsea and Susie were definitely the stars of the show. During the course of the day with all the PR activity and then into the evening with the launch, they had worked with more people than they were likely to meet at one time at a workshop and, of course, they would not normally be faced with so many cameras. I am sure they knew that it was an important day and were as determined as we were to make it a success. One thing that was especially pleasing was watching Susie come into her own. I had always been a little anxious about how she would take to meeting lots of new people as she had always been very quiet and rather shy. But over the course of the practice sessions I had seen her confidence grow and watching her perform for the cameras at the launch I knew she was really enjoying herself – not least because she had stopped looking to me for approval. Now she knew that she could just be herself.

It was getting on for 8.00pm when we finished our demonstration and could finally relax with a glass of wine with our guests. The horses were also very happy to return to their stables and relax with a very welcome haynet. We stayed on to chat to our guests, the last of these finally drifting away at about 9.30pm. As we tucked the horses up, we reassured them that this was not a typical **red horse speaks** day and thanked them for their support. It had been a truly amazing day.

As we were packing up to leave, we were given a video tape of the Grampian TV news item which had been aired as our launch was taking place. We enjoyed watching it when we got home and hoped it would set the tone for all our media coverage.

The following morning, we enjoyed reading the newspapers. I had been worried that **the red horse speaks** might have been perceived as 'New Age' but Liz had done a good job. The reporting was just what we wanted - serious and yet with a twinkle. It was the best possible start. During the morning, we had a phone call from BBC Scotland, one of the few media groups not represented at our press conference. They had obviously realised they had missed a trick and asked us if we would be prepared to do a radio interview for the evening news programme. We agreed to do so and met them at Loanhead. The interview went very smoothly. Then, at the end, the journalist asked if anyone had criticised our work. We replied honestly that we had had no adverse comments and he left, apparently happy with our response. When we listened to the item on the news programme in the evening, we were surprised to hear that the interviewer had then gone to speak to the Scottish Society for the Prevention of Cruelty to Animals to ask their opinion of our work. Happily the SSPCA thought what we were doing was excellent and were very supportive. We also found out that the reporter had also approached our vet practice but they had given him short shrift. Some months later, we learned from a friend that BBC Scotland has a policy of what they described as 'balanced reporting' which to us seemed like introducing controversy wherever possible. In our case, fortunately, it didn't work.

We were touched by the number of emails, cards and phone calls we received from our guests who had clearly enjoyed the event and we also enjoyed sending our thank you's to all those who had helped make it a success. Now we had to turn it into business!

Chapter 15: *Establishing Credibility*

We were pleased that by the time of our launch we already had two clients signed up for workshops so we could immediately follow on from where we'd left off with the practice sessions and the launch itself.

Our first group, scientists from a government laboratory, came for a day's teambuilding. They were amazed at how much learning they had to take back with them. For one of their number, it was her first day with the team. Her post-workshop feedback was interesting. She mentioned that when she went into the office the next day, she felt she knew everyone very well and so it was much easier to ask for help as she found her feet in a new situation. Since then we have had other new starts all of whom have reported the same benefits.

We enjoyed watching how the horses encouraged this quite serious-minded, introverted group to relax and be playful. At the end we invited them to devise a short cabaret which had to involve both horses and all team members. They were wonderfully creative and devised a wedding in which the two horses were 'married' to one another. It had all the attention to detail, however, that one might expect of scientists. They used show jumping poles to mark the aisle while a jump filler served as the altar; they found some artificial flowers to use as decoration and dressed the horses in rugs. There was a choir and some guests for both 'bride' and 'groom'. The bridesmaid and best man led the horses up the aisle. One of the team members, himself recently married, was the minister and did a great job remembering all the words of the ceremony. When the moment came to exchange rings, polo mints were produced – to much laughter! In following up with this group, it was pleasing to know that they had taken this new-found creativity back to work with them.

The next group presented us with our first logistical challenge. They were a team of twenty people and were coming for just half a day with the horses. I would also be working alone with them on the morning before and again the following morning, reflecting on the past year and planning for the year to come. The afternoon with the horses was an opportunity to reflect on their communication as well as to have some fun together.

At that stage, Sue and I didn't feel confident enough to manage everyone together in a single group and so we decided to run two parallel sessions each with two horses. Sue would work with Aidan and two horses whom she knew well and I enlisted an equine massage therapist who knew Chelsea and Susie to work with us. The indoor school at Loanhead is large enough for a 60m by 20m dressage arena with space to spare and so we usually fence off about a third of it with show jumps to make a comfortable working space for our sessions. We worked out that we could divide the arena into three for this large group. That would give us each a space in which to work and an area of 'no man's land' in between. Even so, we weren't sure how the horses would take to having another pair so close by – especially as we had two mares and two geldings. As usual, the horses exceeded our expectation. Once we started to work, the two pairs completely ignored one another and engaged fully with their human students. At one stage one pair were cantering and the other two were completely still, apparently oblivious to all the excitement at the other end of the arena. It worked perfectly and again the clients had fun and learned a lot.

Since then, Sue and I have worked with larger groups, usually with additional horses and facilitators. Our usual ratio is one horse and facilitator for six participants. We have found that in these bigger groups that the facilitators need to have both good horse skills and good business skills as they may find themselves working alone with several participants and a horse. Our largest group to date has been forty people from one of the drilling companies in Aberdeen who came for a two hour 'ice-breaker' session before going on to a two day meeting to discuss challenging operational and personnel issues. They were a very diverse group – ranging from the European Vice President through office and technical staff to the chef on one of the off-shore oil platforms. By dint of careful planning, everyone took part in at least one activity and we also had plenty of observing tasks for those who were not directly involved. The group went off to their meeting in a very positive frame of mind, having noticed that, as one person said, "We have been together for two hours and haven't argued yet!" By all accounts they managed to maintain that for the rest of their time together.

Sue and I were growing in confidence as a partnership, and, to increase our skills and repertoire, we ran several open workshops aimed mainly at women. We specifically chose this strategy as we found that women tend to be more open to new ideas and concepts and so gave us greater opportunities to practice new activities. On a couple of occasions, Liz invited journalists to take part – with the agreement of the other participants – and so the press campaign continued too. Our Ladies Days were great fun and we even had one delightful woman spend her 60th birthday with us – as her gift from her son and daughter. The poor soul confessed she hadn't slept very well the night before, wondering what her children had let her in for but she loved every minute of it, including tea with birthday cake which, of course, also made it extra special for the other participants.

One thing we quickly learned was that however much fun open programmes were, they took a lot more organising as people would say they were coming, then change their mind and so on. It was very much easier to deal with a single corporate client bringing a team of people. Even so, we still enjoy doing two or three open programmes each year and are gradually finding that groups get themselves together to come along.

During this time, we were invited to work with another client group – young people. It arose as a result of a meeting at a networking lunch. At my table was James Martin, who at that time worked for Aberdeen Foyer, an organisation that works mainly with homeless young people and also delivers the Prince's Trust programmes for disadvantaged young people in our area. James asked if we would consider working with young people and he thought that the Prince's Trust groups might benefit from a day with the horses. I told him that it was not something we had done so far but we would be delighted to do so. I had been a teacher and enjoyed working with young people and Sue was also familiar with this age group through Pony Club.

In this work, we are very clear that we are not offering any form of therapy. We offer life skills – leading your life, making decisions, making choices, working together. In fact, we teach them many of the same skills as we teach on our corporate programmes. Naturally, we adapt the context and language and usually find that we need to be more directive in our coaching but otherwise we use the same approach and activities with both client groups.

Our first sessions were a great success and we loved doing them and so they continue to be a part of our work. We have never had to actively sell or market them as the project leaders from the Foyer did that for us. They told all their friends who are involved with the same client group and they found us!

With all the media attention, I had found myself becoming much more comfortable answering questions and also writing articles as I was now being invited to do from time to time. I was also grateful for our stock of photographs. I began to be invited to speak to Rotary Clubs, Women's groups and other similar organisations which I found to be most enjoyable, especially as I had an increasing library of real stories (used anonymously, of course) from workshops to illustrate my talks.

Something I found necessary was to keep practicing and refining how I described **the red horse speaks** so that I had something appropriate to say to various audiences. In one way it seems strange that I would sometimes struggle to describe perfectly something which I had designed and developed. Yet that was simply a fact. It was reassuring to discover that this was completely normal! It's often hard to describe that which we know well – remembering my earlier experience in Canada when I had been asked how to stroke a horse. It's an ongoing challenge too as **the red horse speaks** continues to grow and evolve. I still find myself having a run through in my head or even making some notes before going to meet a prospective client or attend a networking event.

The one aspect of making **the red horse speaks** successful which worried me most was selling. I think this is probably true of many entrepreneurs. After all, if we'd had a talent for selling, that's probably what we would have been doing as a career! Although I'd had my own business for many years, I'd never really had to worry about sales as almost all my work had come to me via word of mouth or was repeat business from satisfied clients. Also, whilst I was involved in very large projects that typically lasted for several years I didn't actually need many customers at any one time. Having worked in the computer industry for some years, I had also been on the receiving end of some very pushy and sometimes almost aggressive sales people – definitely not a reputation I wanted!

One thing that was clear to me, however, was that if I didn't sell any workshops, then I didn't get to work with my horses. Being able to do that was the greatest inspiration of all to find a selling style that worked for me. When I thought about it, it was clear that personal development, whether at an individual level or in a corporate setting, was just that – personal. Mercifully as far as I was concerned, that ruled out any kind of cold calling approach. It would be much more about building relationships and being seen as a trustworthy provider of a high quality service. I hoped that in the end this would lead once again to word of mouth recommendations and repeat business. Once again, synchronicity came to my aid and whilst browsing on Amazon.com, I found a book called "Getting Business to Come to You." Perfect. Just what I wanted! I ordered it at once. It's not a book that comes into the 'light reading' category in any sense of the word as it's a chunky 700 pages long and is packed tight with information. It became my mentor as I developed my selling style and I still consult it from time to time even now.

As I worked my way through it, I was pleased to discover that I was already doing quite a lot that was right and I also found some rather unexpected suggestions – such as volunteering. Of course when I thought about it, it made good sense. As I was building up business, I had some time on my hands and this provided an excellent means of getting to know more people, making myself known, using my business skills and, for me just as importantly, giving something back.

During 2003/4 I was delighted to be able to attend a Common Purpose leadership programme, thanks to a scholarship which made it possible for me to afford the otherwise expensive (for a small business) fees. Common Purpose offers an excellent and rather different approach to leadership in that it brings together leaders from all sectors within a local area. It is as much about building bridges, creating networks and crossing boundaries as learning about leadership. The Charles Handy connection was there again as Common Purpose's founder, Julia Middleton, is featured in Handy's book "The New Alchemists."

Like all Common Purpose programmes, it was based in the community. Although we met for just one day a month, there was plenty of preparation and reading to do in between and so it was actually quite a commitment. The study days were long and intensive and always hosted by a local organisation appropriate to our theme for the day – topics such as Youth Justice, Education, the Local Economy, Health and the Arts. Over the course of the year, my fellow students and I found ourselves at several places we might not normally visit or indeed want to visit. The most interesting, although also the most disturbing, was the day spent at the local prison. It was very sobering to hear all the doors being clanked shut and locked behind us as we made our way to the windowless room which was our base for most of the day. Aberdeen Royal Infirmary which also serves all of the oil platforms out in the North Sea was more comforting. We were able to get a preview of the new children's hospital which was just a few weeks away from opening. On a lighter note, we were able to observe a science lesson in a local school. Over the lunch break on each day, we fitted in two short visits to other local organisations relevant to our theme.

I thought I had quite a good knowledge of public, private and voluntary sectors from having worked in or had clients in all three but even so each study day became a fascinating journey into some hitherto unexplored aspect of the local community. It helped too that we had excellent speakers, many of whom were at the top of their organisation and I soon discovered that almost all the contributors to our days were themselves Common Purpose graduates.

All of the time, one of the key leadership themes was about being a leader in our own community and it reminded me how little I had put back over the years. It was easy to make excuses for myself by saying that my travelling made it hard to commit to anything on a regular basis but there were other ways to make a contribution that could be fitted around my travels. In any case, I wasn't travelling so much and so there were no excuses and I had started to think about what I might be able to offer.

Again synchronicity stepped in and I was invited to be at a judge at the finals of our local Young Enterprise competition. Young Enterprise is a national charity which runs a number of programmes which give school pupils business skills and experience. The competition which I was to judge was the Company Programme which was the culmination of almost a full school year's work for the students involved. They had set up and run an in-school company, sourcing, marketing and selling products, keeping records and essentially doing a bit of everything any company has to do. The final six had been selected on the basis of their company report. I was on the panel judging their presentation but there were others judging their trade stands and interviewing their board members. It was an exciting evening and I was impressed by the standards the students had achieved – and also by the organisation that lay behind it. I have always enjoyed working with young people and there seemed to be opportunities here to offer support.

I volunteered to be a Business Adviser at our local secondary school. Each school which takes part in the Company Programme has an external business adviser from the local community as well as support from school staff. It was always more difficult to find advisers for the country schools and so my offer was gratefully accepted. It wasn't long, however, before my contribution increased as I was also invited to join the board of our local branch of Young Enterprise.

I had a very rewarding year going into school most Friday afternoons and supporting the students as they decided what to make, where to buy the raw materials and how to sell their products. Through my Board duties, I also had links to two other secondary schools whom I also enjoyed visiting. Although I only had time to be an adviser for that one year, my involvement on the Board continued for several more and I eventually served as Vice-Chair for two years before finally retiring when I found myself travelling again. By then, however, I had found some other ways to give back that I could fit in with work and travel commitments. These have included hosting Common Purpose lunch time visits and being a speaker on Programme days. These days I am also involved with their youth programme - Your Turn - and am also a member of the Aberdeenshire Children's Panel and so my commitment to young people continues.

The networking also opened up other opportunities - such as the invitation to give a demonstration of **the red horse speaks** to Aberdeen Entrepreneurs. This organisation usually meets in a hotel once a month for networking and a talk followed by a buffet supper. For their June meeting, however, they came to Loanhead where we had set up the marquee as we had for our launch. It was a very successful event for us and brought us some significant business.

Another opportunity to demonstrate our credibility which came our way during 2005 was to gain recognition from Investors in People (IiP). At that time, they specially wanted to work with small companies and were prepared to offer a discounted rate to make it affordable. At the time, IiP were moving from a standard that was based on business processes to a new standard based more on culture, values and the involvement of all staff in the strategy and decision making. The new standard also placed emphasis on staff training and development. By gaining this recognition, we felt that we would be walking our talk in terms of the importance of people development. We were told that we should be prepared to be assessed against the old standard as very few assessors had been trained in the new one. It was good for us in that it made us review all our processes and update them where necessary. However, we did also have a look at the new standard which seemed to us to be much easier to achieve. This wasn't just because we were so small that inevitably we were all involved but really because we passionately believed that having strong values and a shared culture contributed greatly to business success.

When our assessor arrived, she asked whether we had looked at the new standard. When we (enthusiastically) said we had, she asked if we would like to be assessed using it – and we responded positively with even more enthusiasm! We enjoyed the process and were delighted to receive our accreditation – which also provided a great photo opportunity with the horses and a subsequent news story. It's not every day that businesses have four legged employees on their staff! In the meanwhile, we also had some up to date processes – and that did us no harm either.

The assessment took place at a hectic point in our domestic life as we were about to fulfil another part of our dream and move to a small farm where we could keep the horses. Since coming back from the Secretan Associate programme, part of our vision was having a place where we could bring all our work together. Since then, we had looked at several properties. Some were just not right for us and the one that we really did like was sold for a price well above our bid. (Scotland has a property buying system where would-be purchasers submit sealed bids and so it's always a bit of a lottery. When the market is buoyant, properties can sell for as much as 30% over the original asking price).

Once again synchronicity struck. Since coming back from Australia, Sue had been renting a farmhouse plus a five acre field across the road for her horse. She had built a stable block in the farmyard, constructed in a way that would allow her to move it if necessary. Sue mentioned to her landlord, the owner of a small estate, that should he ever think of selling any of his properties that we would be interested. The following year, he told Sue that he wanted to sell the property she was renting! This was not good news for her but it was perfect for us. In addition to the farmhouse which Sue rented, the package included the steading, a barn and 20 acres of land. We were able to buy the property privately, negotiating the price directly with the owner rather than having to go through the bidding process. It was an ideal solution for us.

Sue was able to buy her field and we were happy for her to leave the stable block in the yard. She lived in a caravan in the yard for some time before moving to a house just a quarter of a mile away. It works well for both of us as it makes sharing horse care very easy.

We were able to sell our house in Ballater very quickly and moved to the farm in June 2005 – just before the longest day. On midsummer's evening, we were blessed with a beautiful sunset and realised how perfectly the house was positioned. Since then we have enjoyed many magical sunsets and sunrises too. Whoever built it back in 1798 knew exactly what they were doing. The only downside was that the fields were covered in ragwort, a plant poisonous to horses. Aidan did a great job in clearing it by hand – the only option at that time of year – so that the horses could join us.

Having the horses at home with us is a source of great joy although I was a little anxious at first. Both of them were so happy and content at Loanhead and I hoped that they would like their new home. I felt like a new mum bringing her first baby home from hospital and was glad to have Sue on hand for questions and support. Chelsea and Susie settled in to their new surroundings very quickly and we soon got used to mucking out!

The next part of our dream is to convert the steading into space for workshops and to build an indoor arena. Sue's dream is to build a house in the corner of her field and move her stable block, thus creating her own yard.

Having the horses at home has allowed us to offer prospective clients a short one hour taster session with the horses as part of the sales cycle. **the red horse speaks** is an experiential learning programme – and so being able to experience it makes it much easier for people to understand how it works and what the benefits are. It's very powerful too as they inevitably go away from the taster with some learning about themselves!

Gradually over the course of 2005, I found myself becoming more confident about talking about and selling **the red horse speaks**. It helped greatly that I knew more people and had started to build a good network. Liz's efforts in establishing a steady drip of PR were now beginning to pay off and so when I mentioned **the red horse speaks**, people would often recall reading or hearing about it. Sometimes too, people had heard about it first hand from one of our clients. The resulting conversations were now much more from a place of interest and curiosity – and that was much better for business. We were no longer 'off the wall' but part of the business landscape!

Chapter 16: *Maintaining our Profile*

Having made a promising start, we now had to maintain our profile and demonstrate that **the red horse speaks** was no passing fad but rather the start of a new way to teach leadership and team skills that would grow and become ever richer in its benefits as we developed it further.

Again we were fortunate in that a solution came to us. We were invited to take part in a programme on International Business Development, sponsored by Scottish Enterprise. It turned out to be one of the best business development courses I have ever attended and was as useful and relevant for doing business at home as it was for international trade.

We met for a day once or twice a month and at each session covered a different aspect of business development. Over the course of the programme we looked at sales, marketing, money, PR and so on from both a national and international perspective. Sometimes the facilitators ran the sessions, sometimes they arranged for speakers to come in and sometimes the group members pooled their expertise and experiences to learn from each other. It was a varied and interesting programme and, of course, we got to know the other participants and their products and services well too. They included an oil services company, a picture framer whose products were targeted at the expatriate market, a wedding dress maker with a large client base in the Middle East, a company who designed beautiful peace jewellery which was made in the Far East and the inventor of the Walkodile, a harness that enabled young children to be taken on walks out of school safely. With all this innovation around me, I suddenly felt much less lonely – and knew I was not the only 'crazy one' in Aberdeenshire! Apart from learning together, we also supported each other as we shared the triumphs and disappointments of growing a new business.

One of the greatest benefits was that I started to look at **the red horse speaks** even more as a 'normal' business which could (and needed to be) planned and analysed using standard business metrics and methods. It was hard work at times but very rewarding as I found new ways of describing the services we offer and the business benefits of our work. As part of this process, I designed a form which we still use to collect feedback from clients. It is structured in a way that enables us to produce overall performance statistics which can be shared with potential clients. Furthermore, we are able to calculate the statistics for each client workshop which is a valuable tool for demonstrating their return on training investment, hopefully leading to more business.

One of the sessions which stood out for me was one about Intellectual Property and it made me realise that all my work in developing **the red horse speaks** had actually created a business asset which I needed to protect but was also there to exploit. I think I had always known that in a vague sort of way and I had, in any case, trademarked the name and our logo but I began to realise that there was more to it. Several people had suggested to me that I should franchise **the red horse speaks** and although I didn't like that exact model, I was keen to find some similar means of expanding the business (of which more later) and then there was the whole area of brand management - so I began to explore these areas further. Yet again I was amazed at where my lovely red horse was taking me and what she inspired me to do!

At that time, I was invited to test out a suite of tools being designed for the Scottish Intellectual Asset Centre in Glasgow. In working my way through the various questionnaires and subsequent analysis, I learned so much about managing both our brand and our business. The results confirmed some of what I knew instinctively – for example, that having the right people working for us would be crucial to our growth – and also uncovered many new things that I wouldn't have thought of for myself. One of the 'surprises' was a useful means of determining the importance of membership of a professional body. While we were gaining credibility, membership was of importance but once we were successful in our own right, it was less so. We still enjoy being members of professional bodies to maintain contact with others in our field but we don't use such membership any longer for credibility purposes. Our brand is sufficiently developed to stand by itself. Another useful tip was about how we used PR. Again, the analysis suggested that PR was helpful in the early days to gain credibility but once we were passed this stage, it was more effective to use it to release news about innovation or about particular success. This was very useful knowledge for setting and allocating the PR budget.

I also began to think about how we could manage our brand more actively and link it into our vision and values. Like any small business, especially one as recognisable as ours, we are our brand all of the time. People, certainly within our community and our professional networks, know our business and see us always as representatives of it. It doesn't matter whether we are 'on duty' at an event or 'off duty' shopping in the local supermarket. How we are is how people will see our brand – and that will influence their decision about whether they want to purchase from us. This realisation brought home to me all the more how important it would be for others who came to work with us as the business expanded to share our vision and values so that representing our brand was easy for them because it is how they are anyway – no pretence necessary.

Exploiting our brand also brought out my creative side and proved to be a good way of meeting one of our performance targets – which is to use environmentally friendly products wherever we can. Once we had finished the supply of chocolate horse's heads which we had ordered for the launch, I began to think about what else we could give people as a 'take-away' from a workshop. Mugs came out on top of the list. To begin with we had used polystyrene cups at workshops. They were convenient but definitely not environmentally friendly and it was this that inspired us to have some branded mugs produced. These are given to participants to use (and keep clean) during workshops and to take away at the end. We also explain that it is part of our effort to be environmentally friendly. We chose a playful message for the mug – 'The red horse has spoken to me.' They are now quite sought after and we enjoy seeing them when we visit our clients' offices.

Since then, we have incorporated more of our values into our performance targets and have thus added more meaning to our brand – so, for example, to increase our efforts to be environmentally friendly, we use local caterers whom we select because they too use local produce, we buy our horse feed from a supplier just a mile down the road and give our muck heap to the local organic farmers who supply us with a vegetable box each week.

Building the brand in this way has been a lot of fun as we see how deeply we can embed our values into all that we do and how we can show people who we are. It has also helped our credibility as we have ensured that **the red horse speaks** has all the metrics and definitions that any good business should have.

As well as developing our service, Sue and I were keen to continue our own learning. We were delighted to be invited by our (by now) friends Lynn Baskfield and Ann Romberg to attend the newly formed Equine Guided Education Association (EGEA)'s first conference in California in 2005 and were even more delighted – and a little surprised – to be invited to present at the second conference. It was our first opportunity to work away from home and we wondered how our work would be viewed. One of the challenges for us (even now) is that we are very rarely able to receive any form of peer review or assessment of our work.

The conference in 2006 set the tone for gatherings which continues even now. Ann is a great organiser and arranges to rent one or more houses near the conference venue which she fills with like-minded people. We were especially delighted to have this support as we prepared for our demonstration. Everything went to plan and we thoroughly enjoyed the opportunity to work with Sadie, a delightful Quarter Horse mare who also seemed to enjoy being with us and the volunteers from the audience who worked with us. The feedback was very positive and left us feeling encouraged. For several years, the conference was a regular event for us and we especially enjoyed the year when we co-presented with Ann and Lynn in a session about partnerships. We began to notice that, good though the conferences were, the house was really the heart of the event for us and occasionally we would find ourselves skipping conference sessions in order to exchange ideas over a glass of wine or during a walk on the beach.

As well as continuing our learning specifically with horses, I also continue to keep up with trends in leadership and personal development, always finding new themes to weave into our work. I also continue to have articles published and speak to local groups as well as presenting at conferences.

The website grew (and still does) as we developed our service and were able to add testimonials. As part of the international business programme, our website was audited and using this information, we were able to make few changes to stay near the top of the lists on Google and other search engine pages. The website easily pays for itself in terms of business generated from enquiries and we also know from our statistics that there are many people who simply enjoy visiting it for inspiration from the photos and quotations that appear randomly on each page. From the beginning, I have enjoyed composing and issuing a regular newsletter and this too has proved to be a great way of staying in touch with clients.

Now we feel we are a normal business and have developed processes and found tools that will enable us to maintain our profile as we continue to grow.

Chapter 17: The Horses

As we were delivering workshops and developing the business, we continued to monitor the horses and still do. We wanted to be sure that they were enjoying all of this as much as we were. Initially the horses lived and worked at Loanhead and so the indoor arena was a place where they were sometimes ridden and sometimes did workshops. This never seemed to cause any confusion for them as they seemed to know the difference according to what I was wearing – and, I am sure, my attitude. Riding gear meant that I was in charge and casual attire meant it was a collaborative effort. This distinction continues on the farm where we work and ride in the same fields. I hope that by setting clear boundaries and expectations, I have made it easy for them to enjoy being both ridden horses and teachers.

Based on the endurance rides I used to do with Chelsea, I estimate that a day's workshop is as demanding as a day's competition in terms of concentration, although the workshops are less physically demanding. Therefore the horses get a complete day off after workshops, just as Chelsea did after a day's endurance. We also try to plan things so that they don't do two or more consecutive full day workshops. Immediately after a workshop, they get some time turned out to graze unless the weather is really bad or it's dark. Some days they amble out and look very tired and I amble into the house, kick my shoes off and pour a glass of wine! Other days, they trot out and these are the days I too am full of joy and feel energised by the work. Some days, we are all tired but also happy with what we have achieved.

I can usually predict how we will feel at the end of a day from the preparatory work I have done with the client. Some clients are inspiring and some totally draining. Dysfunctional corporate teams are generally the most difficult at least initially but once they start to come together, then the energy levels rise and things start to flow.

Groups with learning or physical difficulties and young offenders can often be rewarding, if tiring. The horses are aware of the lack of congruence that is often present in these groups and sometimes that can be hard for them. It's always a balance for us to allow the learning to take place – but not at the expense of the horses. Only once have I deliberately put Chelsea into a situation I knew would be uncomfortable for her because it was the only way that the individual concerned would receive the learning he needed. I knew too that she would be able to tolerate the discomfort for the few moments it would last. The young man concerned was a very talented footballer but also very arrogant and had little regard for his team mates. We invited him to dribble a football around some cones – which he did very stylishly, demonstrating his considerable skill. We then invited him to do so whilst leading Chelsea. She does not like footballs and we knew that she would try to move away from him and that this would upset his style. In the space of a few steps, the young man realised that his attempts to show off were no longer working and he was actually looking rather foolish. To his credit, he started to move the ball very gently, almost at walking pace which calmed Chelsea at once. The learning had worked and he realised that sometimes it takes as much skill and is equally rewarding to keep things simple and thus allow others to join in and be part of his team.

We always enjoy working with good leaders of all ages. We and the horses are inspired by them as they are so quick to recognise the feedback from the horses and then to adapt, expand and refine their leadership style. Chelsea really loves these groups. She is truly at her best teaching master classes in leadership!

Our clients often want to know if the horses enjoy this work – and how I know that they do. I am sure that they do, for a number of reasons:

1. The only time they travel these days is back to Loanhead for workshops and both horses are eager to load into the horsebox and so I assume they are looking forward to the day. Although Chelsea loved to compete, she was never a good traveller to competitions as she was always a bit on her toes. She is noticeably more relaxed travelling to workshops – she rarely sweats up now and even nibbles from her haynet.

2. When I take out the box of accessories which I use at workshops, the horses are immediately interested – rather like picking

up a lead and a dog assuming it is going for a walk. I sometimes work away from home with other horses and I have learned to pack the box while Chelsea and Susie are out in the field as it seems unfair to falsely raise their expectations.

3. We have frequent visitors to our farm – some of whom come for workshops with the horses, others who simply come to see us and enjoy saying 'hello' to the horses too. Whoever it is, the horses always come up to greet them when we go out to see them and are happy to work with them as requested. If they did not enjoy the work, they would probably not be so friendly!

4. During day-long workshops we take a lunch break during which the horses are released into the full area of the school or pasture and given access to food (hay or grass) and water. When we return from lunch, they always come back to join the group ready to start the afternoon session.

We introduce the horses and people to each other at the start of the workshop so that the horses get an opportunity to meet the people with whom they will be working. The humans appreciate this too! At the end of each activity we make sure that the horses get a pat so that they know they have done a good job. In practice people want to do this anyway and they are much more likely to thank the horse than their human team mates – which makes for an interesting talking point.

At the end of the workshop, the horses get patted and fussed over. Clients are then allowed to give them an apple, carrot or other treat – just as I would at the end of a ridden session. If we haven't had the discussion earlier in the day, this is a good opportunity to talk about the difference between bribes, rewards and incentives which makes for an interesting discussion whether in a business or family context! Being rewarded in this way marks the end of the workshop and the horses know they are finished and have done a good job. There is usually a group photo too!

In addition to the usual routine care such as farrier, teeth, flu/tetanus, worming, the horses see an equine chiropractor from time to time. They enjoy Reiki and this seems beneficial when we have been working with difficult clients.

With Chelsea and Susie settled on the farm and me feeling confident about looking after them both at home and in terms of their new role, I gradually began to think about expanding our herd, although with nothing specific in mind. One of the people who attended our launch was Eileen Gillen the manger of Belwade Farm, a rehabilitation centre run by the International League for the Protection of Horses (ILPH) now renamed World Horse Welfare (WHW). She was very interested in what we were doing – especially as it did not involve any riding. Here was a possible future for some of her hard-to-home companion horses. That appealed to me too as it was giving back something to horses just as my volunteering put back something for people. She also indicated that we would be welcome to use Belwade for workshops although it would always be subject to their large barn being free as it was sometimes required as loose housing for new arrivals. Taking clients to WHW works well for all concerned because not only does Belwade have a beautiful situation in the Dee Valley, but we also make a donation to a good cause when we pay for the venue hire. It isn't that I mind paying Loanhead for they always looked after Chelsea and Susie so well but I do also like the idea of supporting an equine charity.

On a visit to Belwade prior to taking some clients there, we were walking round the paddocks to see which horses might be suitable for our workshop. We entered one of the paddock and almost at once a sturdy grey pony came up to greet me. He was very shy so I stood quietly and allowed him to sniff me and then I was gently able to stroke him for a moment until he stepped away again.

"Would you like to adopt Storm?" enquired Eileen. "He seems to like you."

"Yes," I said, "although he'll have to wait until we get all of the fencing tidied up."

We had done the essentials for Chelsea and Susie but there was more to be done in the other fields to make them stock proof and I wasn't sure how long it would take us to get everything into place. I added that if she managed to find a home for him before we were ready that she should let him go as that was really the priority for him. Although I wanted to extend our herd, there was no urgency.

When we went to do our client workshop, I enquired after Storm. I learned that he had indeed been adopted by someone else and, whilst I was happy for him, I found myself feeling incredibly sad that he would not be coming to stay with us. Several weeks later, however, Eileen phoned me to say that Storm had come back. The placement had not worked and was I still interested? Of course I was! By this time, we were making hay and the mares were in the only area left for grazing but a few more weeks and we would be ready.

In the meantime, I started to visit Storm from time to time just so that he got to know me a little better. I learned that he had had a tough time prior to being rescued. He had run wild as a stallion till he was six years old and in that time had also been beaten, chased and generally harshly treated. Despite the best efforts of the staff at Belwade, it was clear that Storm was always going to be too flighty to be safe to ride and so he was classed as a companion horse. Hearing his story, I felt even more touched that he had chosen to trust me even in such a small way. During my visits, he became a little more confident and I was sure he would one day be a valuable member of our team – but I was in no hurry as I knew that he would take time to settle with us, especially following the earlier unsuccessful adoption.

Of course we couldn't just have one gelding living by himself and so I enquired if there might be another suitable one to keep him company. As I knew that Storm would be quite a challenge, I requested that the other horse be more straightforward. Eileen immediately recommended Darcy – a very handsome and very large Cleveland Bay x Thoroughbred. He came to WHW with behaviour problems, often rearing when being ridden. It was eventually discovered that he had had a trailer partition collapse on his back as a two year old. As with Storm, everything was done to rehabilitate him but it seemed that although his back was fine, there was still some scar tissue that means that the weight of a saddle and rider causes him discomfort. Thus Darcy became Storm's companion horse.

We were delighted to have them with us and started to get to know them. Darcy was and still is a gentle giant, standing just over 17 hands high. He adores people and can be a little pushy at times although is very quick to move back when requested. This makes him ideal for teaching people how to maintain boundaries. His size too is very useful as many people find him intimidating – and that can also be very helpful when dealing with anyone too big for his/her boots! On the other hand, he adores small children and will allow them to lead him round the field, content to walk very slowly and keeping his head down so that his eyes are level with theirs. It is always a most endearing sight.

Not surprisingly, Storm was very timid at first in his new surroundings and wouldn't be caught. He was terrified of lead ropes – probably because he had been hit with them. This didn't concern us as we knew that he would follow Darcy if we needed to bring him inside for any reason and Darcy was always willing to be caught. We decided to give them a small feed every morning – not so much because they needed it but more to give us the opportunity of handling them in a positive context. We discovered that Storm had probably been bribed with food for, unlike normal horses, he wouldn't come near us when we went into the field with the feed buckets. For a while, we had to set his on the ground and then allow him to come up and eat.

It wasn't long, however, before he was happy to approach us when we went into the field and eventually he would even arrive before Darcy, a sign of his growing trust and confidence. I like to think that Chelsea, Susie and Darcy were all good role models for him as they are always pleased to see us and so Storm began to realise that this was somewhere he would be safe and loved. From the start, we encouraged our friends to visit our new arrivals (as well as Chelsea and Susie, of course!) explaining that Storm would like to be friendly but would also be a little nervous. Gradually he became more confident and these days he enjoys meeting people although at times, he can still be a little anxious. He too is an amazing teaching horse, always gentle with those who need his love and also very effective at helping those with high energy to see their (often unintended) impact on quieter souls.

He has been a great teacher to us too. The staff at WHW recommended we always keep a headcollar on him as he disliked having it taken on and off. We did this but discovered that it wasn't much help since he would drag us off our feet if we attempted to attach a lead rope. It was clearly a flight reaction from a place of fear and so we decided it was not going to be helpful for any of us to allow this to continue. There were several times when we found him with the headcollar pulled over one ear which was not comfortable for him. On these occasions we brought him in by allowing him to follow Darcy and then would have to undo the headcollar and put it on again. This did nothing to make putting it on and off a good experience for him as we always had to squeeze it against his brow to unbuckle it or pull the whole thing over his ears. In the confined space of our courtyard, we could always corner him to do this but we noticed that when we did so, he would just shut down, give up and allow us to do what we needed to do. It wasn't the relationship we wanted with him.

In the end, we stopped putting on the headcollar and simply forgot about it for a while. We found that he would stand with us in the field to be brushed and would allow us to pick up his front feet. Moving behind him was always more difficult as he liked to see where we were. Clearly people had chased him from behind. We also had to move our hands slowly as a quickly raised arm would always cause him to flee, confirming to us that he had indeed been mistreated. We wondered how we would deal with the farrier's visits as it was important to keep his feet in good shape but in fact it has not been a problem. None of our horses wear shoes so only require a routine trim and rasp and Storm will stand perfectly for the farrier without being restrained. It just takes someone to stand at his head to reassure him and give him the occasional treat.

The first annual teeth, flu and tetanus vet visit was more of a challenge. Fortunately our vet has had experience with wild animals so Storm got the routine care he needed. By the next year, however, he was a normal horse. I was able to gently slip on a headcollar and a lead rope and he was treated in exactly the same way as the other horses. We hadn't practiced with him or made any special preparations other than spending time with him and standing with him for the farrier. It was a victory for 'doing nothing', for calmness and patience. He has reinforced the gifts of patience and consistency that Chelsea gave me years before and, between them, they ensure that I get plenty of practice!

More recently, we have added Bluebell and Katie, a mare and her foal, to our herd. My plan had been to leave them in peace and not work with either of them for a while. They, however, had other ideas. At the first event after they arrived, we were working outside and they insisted on coming up to greet the participants who, of course, regarded being so close to a foal as a very special experience. We quickly decided it was more practical to allow Bluebell and Katie to do a little bit of work with us rather than try to exclude them altogether. Katie is very calm and has very good manners – thanks as much to her mum as to us – and so they have become a great asset, allowing us to have very good conversations about house rules, respect, discipline and consistency. In return, Katie is meeting a variety of people which is good for her development. Having four horses turned out together increases the dynamics and relationships and so herd observation is also much more interesting for us and for our clients.

As our work expanded, we began to work in different parts of UK and so had to find other venues and horses. Sue's experience of competing on the national eventing and carriage driving circuits was very useful in that she knows venues and horse people in most part of the country. Wherever possible we like to use venues that belong to charitable organisations such as Riding for the Disabled (always a first choice if we have any disabled participants) or World Horse Welfare so that we are giving a donation to a good cause. However, it is always of greater priority to find somewhere that is both safe and convenient for our clients.

Most people are delighted to have us use their premises. Very often we can arrange to use them at 'off peak' times and thus provide some extra income. Before we use a venue for the first time with clients, we are happy to give a taster session to the owner/manager and staff so that they understand what we do and how we do it – although the need to do this is diminishing as people become more familiar with our work. Indeed, these days we are often invited to use venues!

Chapter 18: The Red Horse Academy

During the International Business Course, I began to think about how to grow the business and how to take it overseas. Having lived in South Africa for a number of years and then followed its progress from apartheid to the Rainbow Nation, I would love to be able to help bring people together there. Sue has experience of Australia and so developing our business into these countries seemed to be a possibility.

First of all, however, we needed to have more people! Even before we had international ambitions, we hoped that the time would come when we needed more people to deliver **the red horse speaks** in UK. I knew from the work I had done with the intellectual asset tools that having the right people was critical to our success and I had also realised that we would need to train them ourselves so that we could be as sure as possible that they would deliver the workshops to our standards and in a way that honoured our brand. However, we hadn't got much beyond talking about it until we went out to EGEA in 2006, the year we demonstrated our work. All we had decided was that it would be called **the red horse academy**.

One of the other people staying in our house during the 2006 conference was Mercedes Jiminez who had come from Madrid. She had contacted Ann Romberg and asked about the conference. Ann immediately invited her to stay with us and we were delighted to meet a fellow European with an interest in learning with horses. Mercedes' original plan was to come back to US to learn how to deliver this work but after seeing our demonstration she asked us if we taught others. We told her that we didn't at that time but it was something we planned to do in the not too distant future. Mercedes was just the spur we needed. After the conference, she emailed me regularly asking how our plans were going.

The curriculum was developed thanks in great part to an unexplained error on our website (another piece of synchronicity). The previous year we had advertised for some people to pilot a new programme we were planning to offer. For some reason, the same page appeared on our website early in 2006, albeit showing the 2005 dates. Two people, Joy Wootten and Sue Stephenson, both from Aberdeen University, emailed me to ask if they could take part in the pilot. I felt so bad about the error that I invited them out for a short session with the horses one afternoon. It was rather like meeting Lynn and Ann and we were soon chatting away as if we had known each other for a long time. I was glad they had not spotted the 2005 date! By the end of the afternoon, Joy and Sue were so interested and enthusiastic that they wanted to be able to help in some way. I explained about our plans for **the red horse academy** and they at once volunteered to come and spend a day designing the programme.

What a gift from two ideal people! Joy is involved in staff development at the university and has designed many very successful courses. Sue worked in the accounts department and has a sensible approach to all things financial as well as knowing a bit about learning – and she is a long time horsewoman! We planned the date and ended up having a wonderful day together which saw our initial ideas discussed in enough detail to be developed into a full course of study.

We were finally able to tell Mercedes that we did indeed have a programme and so she spent a week with us in July 2006, testing out our curriculum and giving us invaluable feedback. Her copious notes lovingly translated into English were a very generous contribution to our manual.

Alongside developing the curriculum, I had to design an Associate agreement. I wanted it to reflect the same values as our application process and **the red horse academy** programme itself – and, of course, it needed to contain the appropriate legal details of how the agreement works and what it covers. Several years ago, I had been introduced to the "State of Grace Document" devised by Maureen McCarthy and had used it as the basis for agreements when working in collaboration with others. We devised our own State of Grace statement as a preface to our associate agreement and we hope this gives a context to the necessary legal paragraphs which follow. Although the agreement is still work in progress, it is working well so far and the State of Grace preface has not only been appreciated but has inspired others to use it in their own working agreements.

It quickly became clear that one of the challenges for people entering the world of learning with horses was a lack of business skills and so we have added a module to our training course to give our students basic business skills. We have concentrated on selling which appears to be the biggest gap – and also the one that causes most anxiety. I was delighted to collaborate with Annette Brooks-Rooney to produce our "Sales Bible" which is available to all our students.

I gradually began to do some detailed thinking about who I wanted to attract to **the red horse academy** and how I was going to find them. From my own experience of learning to do this work, I decided that initially at least, I wanted to grow a community of highly skilled practitioners rather than simply trying to put as many people as possible through our programme and then waving them goodbye at the end of their time with us. Besides which, I knew that by now **the red horse speaks** had brand value in that it would be much easier for people to be able to use our name rather than having to start their EAL business from scratch. It would be important, however, to find the right people – another point highlighted in the intellectual assets exercise.

I designed a clear application form and process which I hope is fair both for me and the applicant and we are attracting an excellent team. I am happy for this to grow gradually and to allow each person the best possible opportunity to develop their **red horse speaks** business before accepting further applications from people in the same geographical area. One of the joys of our community is the range of talents that we share between us and can call on for the benefit of our clients.

I anticipate that as our community grows, I will add on 'post-graduate' programmes and other courses that might also be attractive to people from outside our community, some run by us and some by invited teachers from elsewhere. I have benefitted from learning with people such as Barbara Rector who is one of the original founders of learning and therapy with horses and Robin Gates, whose horse skills sit perfectly with our work. I hope that people such as Barbara and Robin will come and teach here in Scotland. Indeed as I write this, Barbara will be doing just that in April 2009.

The other gap which Sue and I have identified is a need for a defined level of skills for the coach/facilitator in the partnership. We had always had a clear idea of the skills required by the horse person and covered them in our programme but we began to realise that the business person also needs to have specific horse skills. So we have set about developing a module which will be ready to test out very soon.

All of this is still at an early stage but I enjoy teaching and supporting others and so I look forward to developing this aspect of our work.

Part III: *Science*

Chapter 19: *"What We Need is Research"*

As I began to develop **the red horse speaks**, I was aware that the feedback we were getting was very positive. I was also aware that neither I nor our participants really knew why the workshops were so good. As I attended more conferences, it became clear that no-one else knew either. We all had plenty of anecdotal evidence – and there was even more from people who had been offering therapy with horses for many more years.

As with much else about **the red horse speaks**, my decision to do research came about more by synchronicity than conscious planning. Early in 2006, I was approached by Audrey Hendry from the Education Department at Aberdeen University who asked if I would be prepared to host a student for three days of 'industry experience'. The student turned out to be a local deputy head teacher who was on a programme sponsored by the local authority for those who were seen as future head teachers. She loved horses, having several of her own, and her particular interest was in communication and so spending time with us brought all her interest together very neatly. We thoroughly enjoyed having her with us and shared her joy on the final day when she arrived late and very excited because her mare had foaled during the night and produced a healthy colt.

During the planning and follow up involved in this visit, I casually mentioned to Audrey that it would be interesting to do some research around our work with the horses but that I had no idea where to start. She immediately suggested we have lunch together - and so we did. Audrey very patiently explained the various routes I might follow and the difference between taught post-graduate programmes, where students follow a programme of lectures as well as completing a dissertation, and research programmes, where the emphasis is on original work leading to a (usually longer) dissertation. She also explained the difference between quantitative and qualitative methods and suggested some books which she hoped would help me begin to understand more about these different approaches.

I have always loved reading books - especially non-fiction - and so the prospect of reading more was a positive delight. Even as a small child, the adult section of our local library was much more appealing with its books about real places and people. The only exception, as I mentioned earlier, were stories about horses! Now with the internet and services such as Google Scholar it felt like having all the world at my fingertips in my quest for knowledge.

One of the approaches which Audrey suggested was Narrative Enquiry, a form of investigation which is based on analysing stories and experiences. Collecting stories about people's experiences with the horses would easy as we did it already and so I was eager to find out more about this method. One of my early Google searches threw up a whole PhD dissertation based on Narrative Enquiry. It was by a student at Bristol University, named David Quinlan, and was based on his research into work in the field of mental health. Another pivotal moment! I'm not an expert in mental health but knew enough about it for this dissertation to be very readable, especially as it was more about people who work in mental health than about specific conditions. Whilst not wishing to imply that this work was simplistic in any way, I finished reading it and, for the first time, thought to myself "I can do this!" I have never been able to contact David Quinlan but I hope he feels the gratitude I frequently send his way.

Having decided I wanted to research for a PhD, I needed to find an academic institution at which to study. My first thought, of course, was to contact the two local universities - Aberdeen and Robert Gordon. I thought that they might be interested in supporting the work and research being done by a unique local business - but both rejected me. It's somewhat ironic that three years later, I was invited by Aberdeen University to speak to their interns in Entrepreneurship about how to develop and grow an innovative business. Shortly afterwards, Robert Gordon's asked if they might film me for a Vodcast which will be used in a programme they are offering to small and medium sized businesses. Both have long since been forgiven as, in retrospect, my applications were probably too long on enthusiasm and too short on credible academic content.

Soon after, however, Sue Stephenson by now one of our **red horse speaks** associates, put me in touch with Dr Carol Hall of the Equine Science Department at Nottingham Trent University (NTU). The Equine Science Department is part of the School of Animal, Rural and Environmental Studies (ARES) and is based at Brackenhurst, near Southall – so quite separate from the other two campuses of the university which are both in the city of Nottingham. When I contacted Carol, she was immediately very enthusiastic as she was aware of learning with horses but had little direct experience of it although the department did offer a module in the Therapeutic Use of Horses. I was invited to lecture to students taking that module and to do a demonstration of our work.

Sue Hendry came with me for the demo but even before we got there, we ran into problems with the school's Health and Safety policy which demanded that any time a horse was handled, the handler had to be wearing a hard hat, boots and gloves – which, of course, we never do in the normal course of **the red horse speaks** or, indeed, while handling our own horses at home. By the time we arrived in Nottingham, we were told that we, as external facilitators, had been given dispensation and didn't have to wear hats, boots and gloves but the students participating in our demo would have to follow school policy.

It was interesting to see these students, supposedly being trained for work with horses, look so uncomfortable when simply being on the ground with loose horses. We also discovered that they were never allowed to lead more than one horse at a time. These 'rules' had struck Sue even more and it kept us talking most of the way home. How would these students manage to bring in several horses at a time – as most of us who own more than one horse do daily and which would be routinely expected of the students were they to work on a commercial yard? Just as importantly, what about the horses? They were being treated almost like wild animals. How did they feel about the lack of affection? Whilst their physical needs were being well met, we couldn't help but worry about their emotional needs.

That said, there was interest in our work and my research and so I continued to talk with Carol and made some more visits to NTU to see how my research might fit in. Carol alone was unable to supervise me and in any case, the scope of my study was beyond simply equine science although it was decided that the Equine Science School should be my base. It was quite a revelation to discover just how complicated it was for the university to have a student whose interest spanned more than one department! The amount of paperwork was also quite a surprise – but more of that later.

I shared my plans for research at EGEA in January of 2007 and once more my good friends Ann and Lynn stepped forward and offered to help in any way they could. The model of sharing the house and learning from each other inspired us to meet up in Minneapolis later in the year for what was to become the forerunner for the Equine Guided Coaches Circle, of which more later.

Ann and Lynn were brilliant. They were so eager to help and to know what I wanted to achieve and so we had several conference calls to plan our event. Although I wanted to look at why learning with horse was so effective I didn't yet have a detailed study plan and so I asked if we might just meet in a spirit of curiosity and see what happened. They knew exactly what I meant and arranged a wonderful weekend with a delightful group of fellow enthusiasts from their local area. The location was to be Healing Arts Wellness Centre (HAWC) near Hudson, Wisconsin, owned by Su Wahl. I met Su at EGEA in 2007 as she had come along with Lynn and Ann and was one of our house group. She is a naturopathic doctor and introduces work with horses to her clients as and when appropriate. I loved her enthusiasm and energy – and so being based with her was perfect.

While we were together at EGEA, Su had mentioned that she had just purchased a new computer system for doing Limbic Stress Assessment (LSA) which she was using to measure overall physical and emotional well being in her clients. Su's new system was also able to select and allocate remedies and she was sure that the horses could be set up as remedies on the system by using mane and tail hair to determine their electromagnetic signature – which was what the system used as a basis for holding all the naturopathic and homeopathic remedies already set up on the computer.

In talking about my research plans, Barbara Rector also stepped forward and mentioned some work that one of her Adventures in Awareness colleagues, Lisa Walters, had done with others to measure the impact of humans on horses. Barbara put Lisa and me in touch and so my September trip was extended to include a visit to Lisa in California as well as to Ann, Lynn and the other in the Midwest.

The equipment which Lisa had used came from the Heartmath Institute and uses heart rate variability (HRV). HRV indicates the regularity and evenness of the individual's heartbeat as well as the number of beats per minute. Having very regular and even heartbeats is described as being in a state of psycho-physiological coherence. Put simply, reaching a state of coherence requires warm, positive thoughts felt in one's heart along with calm, regular breathing. I was able to purchase a Heartmath monitor from Hunter Kane, Heartmath's associate company in UK, and they gave me permission to use it for my research.

We now had a broad agenda for our study weekend. I would share my research thoughts to date and we would try out LSA and Heartmath® with the horses.

In between planning this, I was still working out the focus for my research. I began to see that there seemed to be four strands to the answer to my question about what makes learning with horses so effective:

- The nature of qualities that are required today to be successful whether as an individual, as a team member or as a leader

- The way we can most effectively learn these skills

- The nature of horses themselves

- The horse-human interface and interaction during a workshop

I presented this, along with my plans for experimenting with LSA and the Heartmath® monitor to Carol before the summer holidays. She was interested in the Heartmath® equipment as I was able to take it down to NTU and allow her to try it. However, she was much more sceptical about LSA – not helped by my lack of detailed knowledge of exactly how it worked! I was by now used to scepticism with regard to **the red horse speaks** so didn't worry too much about this and looked forward to my study trip to America.

The trip was more than I could ever have dreamed possible. For most of the study time, I stayed with Ann in Minneapolis and we travelled out to Hudson each day. Spending time with Ann was a treat in itself. We had fun discovering just what similar paths our lives had taken and that we had been born just three days apart in June 1951.

When we arrived at Hawk's Ridge on the first morning, we were greeted by Su who told us that the herd of working cow horses belonging to her husband had broken out of their field and were waiting for us in the barn! This herd had not so far taken part in any of Su's work as they were often away competing on the college rodeo circuit but they seemed determined to join us – and so we decided that it might be fun to work with them. In any case, it might be useful from a research point of view to work with horses with no previous experience of what were planning to do.

Su started us off by introducing us to Limbic Stress Assessment and explaining how it worked. Her computer system is actually the most recent in a long line of developments in the field of electro-dermal testing which goes back as far as the ancient Greeks. Electro–dermal testing in its modern form was discovered by Reinhold Voll in the 1950's when he developed an electronic testing device which used electricity to find acupuncture points. He demonstrated that these points have a different resistance to a tiny electrical current passed through the body when compared to adjacent tissue. Voll was successful in identifying many acupuncture points relating to the diagnosis and treatment of specific conditions, including lung cancer, musculoskeletal injuries and allergies.

The LSA system measures stressors in the body using Galvanic Skin Response (GSR) which Carl Jung established could track physiological arousal or stress in the body in the early 1900's. The measurements are taken using a hand cradle which records the electromagnetic frequencies in the meridian lines of the body and uses this energy as information about the state of balance of the body. The output is a diagram showing the extent to which the meridians (and the organs and glands they affect) are in or out of balance. As described above, the system can also allocate one or more remedies. In this case, we restricted the choice of remedy only to the horses.

For our first LSA session, we had readings taken but were not told which horse we were matched with as a remedy. We moved into the barn and Su led us through a gentle facilitation before leaving us to our thoughts as we mingled with the horses. I interacted with several of them and then went to sit on the ground slightly away from most of the herd. One of the dogs came with me and invited me to throw a stone for him. It was fun to play with him whilst quietly reflecting. Then I was aware of Gyp, one of the horses, coming over beside me. I acknowledged him but continued to play with the dog. Gyp pawed the ground and so I spoke to him, albeit in a rather off-hand manner, as I threw another stone for the dog. Then he pawed my arm very intentionally with his hoof. It wasn't sore but the message was clear. I needed to give him my attention. I stood up and stroked him for a little while – and then he walked off.

As we gathered again to talk about our experiences, Su told us the horses we had been matched with by the LSA system – and mine was Gyp. He clearly knew that I needed him even if I didn't. It was to be the first of many amazing interactions with horses.

I was leading proceedings the following day and suggested that we go and sit out in the airy barn with Su's regular teaching horses.

We took our chairs out and sat in a circle. I explained what I had done so far and about the four main pillars of my research. The horses wandered around and gave us great feedback. When we were talking about how we engaged with them as part of our work, they closed in a circle around us, standing behind our chairs. In facilitation terms, they held the space for our conversations. When we got too theoretical, they moved away and at one point, where we had moved so far off topic, they all began to pee! We had an excellent discussion and valued the feedback from our four legged partners. In fact during that morning, not only did we have the horses with us but also two dogs, several cats and a flock of guinea fowl, all apparently attracted by the energy we were generating.

In the afternoon, we worked with the Heartmath® equipment. Our initial experiment was very simple. Sitting inside, we linked each person to the PC based Heartmath® programme using a finger-pulse sensor and noted how long it took each person to reach a state of coherence and for how long they could sustain it. None of the participants had used the device before. It was noticeable that those people who included some form of meditation in their daily practice definitely found it easier to reach and sustain coherence – in one case even when a telephone rang very loudly in the room we were using.

We then moved outside into the pasture with the horses and repeated the exercise with the portable Heartmath® device while the person being monitored was stroking or standing near the horses. We didn't do any set activities. We simply let people be with the horses for about half an hour.

In all cases, people reached coherence and sustained it for longer in the presence of the horses. This very simple study not only gave us some measurements, it did appear to indicate that just being with horses was good for our volunteers. Clearly the activities many of us use when working with people and horses can provide additional and valuable learning but it would seem that just being with horses also has a value and perhaps contributes to our readiness to learn.

There was one interesting reading, however, which is worthy of further investigation. Several people walked around the horses while they were stroking them and, in all cases, there was a momentary loss of coherence whilst walking round the back of the horse. This was surprising as no one was afraid to do so or felt any anxiety – and yet something was clearly triggered by this action.

This was a very small scale initial preliminary study but the results are sufficiently encouraging for us to design a more comprehensive study specifically to examine the hypothesis that being with horses is an effective means of achieving coherence and thus wellbeing, a brain state associated with increased capacity to learn. The method itself needs to be refined as there was an element of 'performance anxiety' at the start of the test. We also need to correlate the PC and portable monitors more precisely to ensure that the timings are completely accurate. A control group would also help show if the coherence readings were due to time spent with horses or to continued practice without horses present.

As well as checking out ourselves, we attached portable monitor to the ear of one of the horses to check out her coherence. I had tried this with my horses at home and so knew that it could be done safely. It was no surprise either at home or in Wisconsin to know that the horses we checked appear to live in a state of coherence. Even when I tried it with Chelsea, she quickly moved to a state of coherence, despite her unfamiliarity with the equipment.

That evening we stayed with Su spending the night at her house by the St Croix River. Ann and Lynn had arranged for their friend Bob Kokott, organic chef extraordinaire, to cook a gourmet dinner for us as we discussed the events of the day. It had been a beautiful warm day and so later on we decided to go down to the river. As the evening cooled, we built a fire and continued to talk. The 2008 presidential election was starting to appear on the horizon and we wondered who the candidates would be and what the outcome would be. Watching the Inauguration Speech more than a year later I couldn't help remembering our conversation under the stars with the river lapping gently in the background.

The following day, we worked again with LSA. Each participant was given an LSA and the results recorded for overall stress levels and the top three most stressed organs. The system also allocated a horse as remedy and this time we were told who our horse 'remedy' was. We spent about 30 minutes, stroking and interacting with our allocated horse. No specific activities were recommended and all interactions took place on the ground. Immediately afterwards, a second LSA was administered. The results were noted and compared with the first assessment.

There was a noticeable, quantifiable reduction in overall stress for all participants and in the three most stressed meridians. My own results are shown below.

Measure	*Before Horse Remedy*	*After Horse Remedy*
Articular Degeneration	56	0
Thymus (Immune System)	65	0
Pineal Gland	52	21

The results amazed me. I do suffer from arthritis (a form of articular degeneration) and it had been quite painful when the initial tests were done. I was also very tired (immune system,) and was still suffering from jetlag (pineal gland) after my flight to USA. Following the horse remedy, the pain in my fingers had reduced considerably and I felt less tired. The others were equally impressed by the accuracy of their results. As with the Heartmath® experiments, we needed to tighten up the protocols but it was a good start.

It had been a magical weekend and I was grateful to all the humans and animals who had provided such inspiration. I finished my visit by spending a couple of days with Lynn during which I was able to ride her lovely horse Baroness. I also experienced a Midwest thunderstorm on my last evening. We planned to go out to dinner, visiting a local tack shop on the way. Schatzlein Saddlery was celebrating its centenary and as well as all the modern tack and clothing required by both horse and rider, it had a great display of old photos and equipment going back to its earliest days. On our way into town, we had been listening to the road reports about the thunderstorms in the area and whilst we were in the shop, the heavens opened in our part of town and within a few minutes the street outside had become a lake. We eventually managed to get back to the car and on to the restaurant by which time the storm had eased. We finished the evening by having a wonderful night time tour of the city courtesy of Lynn's partner, Bill. Then it was time to go to California to meet Lisa Walters.

I had met Lisa only very briefly at the very first EGEA conference but was looking forward very much to meeting her again and hearing about the work she had done with Heartmath along with Dr Ellen Gerkhe. As well as talking about Heartmath, Lisa had also invited me to a two day programme being run by her good friend Robin Gates. I had seen Robin and her mentor Carolyn Resnick at EGEA the previous year and had been so inspired by the beautiful way in which they worked with horses that I had bought Carolyn's book 'Naked Liberty' in order to learn more. I was so delighted to have the opportunity to attend the workshop.

Lisa and her partner Richard were wonderfully kind hosts. My plane was delayed due to the last of the thunderstorms in Minneapolis and so the airport bus I eventually caught in San Francisco made slow progress north to Santa Rosa in the Friday afternoon traffic. I phoned Lisa who told me not to worry and arranged for Richard to collect me from the bus station and take me to their home for dinner. It was such an enjoyable weekend, especially as I met up with a few more old friends who had also come to Robin's workshop. More important for me personally, I learned a lot from Robin that I felt would help with Storm whom we had had for about a year by then. It was also reassuring to know from Robin and her experience with wild Mustangs that we were on the right lines with what we had done with him so far.

Lisa was very generous in sharing her work with Ellen and I left feeling even more positive about my research. By the time I had left, I had also planned to spend more time with Robin the following January as she had agreed to do a private workshop for Ann, Lynn, Su and I before EGEA.

After a few days with friends near Santa Cruz, I flew home is high spirits. For the first time ever, we believe, we had measured the impact of horses on people. The experiments weren't perfect but we certainly had something to go on. I was looking forward to sharing this with my supervision team at NTU.

Chapter 20: A Year in the Academic Wilderness

In October, I went down to NTU, excited about our achievements. I had prepared a presentation for my supervision team but it was received with surprisingly little enthusiasm with more emphasis being placed on their scepticism about LSA and Heartmath®. I decided not to be too disappointed and began to look at how I could design my overall project. Shortly afterwards, I was asked to formally register with the university as soon as possible. The Equine Science Department was being subjected to a Research Assessment Exercise (essentially an external inspection) and it was seemed that having me registered as a student would be helpful for this process.

Registering was the first inkling I had about the paper chase I would need to follow. There were forms for everything – and mostly each form needed the same information. It reminded me of work I had done in the health service years before when we had some patients count the number of times they needed to give basic information about themselves and their local doctor. There were also a series of deadlines for submitting progress reports and finally a project approval form to register my PhD which I was required to do within twelve months. The idea of reporting progress was quite reasonable but the forms were very dull. I began to realise that academic research at this institution anyway, was not really about innovation but about following a process and being able tick boxes. I was still very enthusiastic about what I was doing and one thing I did enjoy about my visits to NTU was having a few days where I could concentrate completely on studying.

One thing that I did not enjoy was the state of the horses. They all looked sad and shut down as I watched them being moved about the yard. I never saw a student or member of staff pat a horse or show any affection. The horses were in good physical condition but I felt for them, knowing how much our horses at home thrive on the emotional and spiritual connection between all of us. I always tried to arrive at the equestrian centre with some time to spare and spent it sending some healing to all of them. I still think of them and send them energy and hope that one day they will be with caring, loving owners.

I soon discovered that I could stay in a student house during my visits. Each 'house' consisted of six en-suite study bedrooms and a communal sitting/dining room with a kitchen area. The houses, which were single sex, were in blocks of four and had all been recently built. It was all very different from the hall in which I had lived in St Andrews many years before. I had the pleasure of sharing with four young women who made me feel very welcome indeed, even inviting me to join them for a birthday celebration.

Life was not without its excitement, however. About one o'clock one morning I stirred, hearing some-one come back into the house. Then there was a rumbling from the en-suite unit in the corner of my room. My first thought was that there was air in the pipes but the noise got louder and the whole unit started to shake. Then I heard shrieking and running feet in the hallway. I got up and went out. There was a little gathering in the room opposite mine, already watching the BBC News Channel and logged on to the internet. What we had felt was an earthquake! One of the students was a geographer and was already on to the British Geological Society website as she knew they kept a live record of seismic activity. Sure enough there was the graph showing the activity we had felt. The epicentre was some miles to the north of us but it had been a significant quake. Soon everyone was phoning and texting to make sure that local friends and family were safe and the BBC news was already broadcasting texts and emails from people more directly affected. It was an interesting example of how quickly news and information spreads these days. Within about ten or fifteen minutes of the event, we knew all about it. We continued to watch the news for little while longer before going back to bed.

Once I had formally registered as a student, I discovered that I had to do a Research Methods Module as I had not done this as any part of previous study. Fortunately, it was possible for me to join some other equine science students who were also distance learners. The module was delivered in a series of two or three day blocks which suited all of us. Although the module was interesting it was hard work as I have never done statistics. Most of my class mates were much more expert so I found it quite hard to keep up with the pace during lectures but given time on my own I could usually work out what I needed to know. It wasn't actually so different from the logic I had enjoyed studying so many years before. As well as statistics, the module also covered topics such as literature reviews, scientific writing and presentation skills. There were various assignments to be completed by the end of the year.

As well as studying, back home I was still busy with workshops. One of the most interesting that we did was also in an academic setting. Craig Brooks-Rooney, son of our red horse associate Annette, was in his final year at Cambridge and co-organiser of a field trip to Chile for fellow final year students. During the summer he had asked if we would help sponsor their trip and of course I said yes. He suggested that part of our sponsorship could be in the form of a **red horse speaks** workshop for the field trip team. We were delighted to do this and started to plan the logistics.

One of the students in the group was a member of the university riding club and this enabled us to use the equestrian facilities at the university vet school. Sue Hendry made contact to explain what we would need and, in her usual way, struck up a good relationship with Alison Schwabe, who runs the equestrian facilities. After our experience with NTU, Sue asked about hats, boots and gloves and was very relieved when Alison said that that was not necessary – basically, because vets don't wear them!

We had a great day with the students in spite of the wet, cold Fenland weather that chilled all of us to the bone. Sue and I decided that as the students were going to a strange land and only one of them spoke Spanish that we wouldn't go and check out the horses in advance as we would usually have done. Sue had found out enough about them from Alison to know that they would be fine for us to work with, especially as they were used to vet students handling them. Instead we took the students with us at the start of our day together and involved them in our observation and discussion. By the time we had taken the horses into the arena, the students already had some practice in going into an unfamiliar 'land' where they didn't speak the language! The vet school equestrian centre was a comfortable and relaxed venue in which to work and the students were quick to settle and participate.

As well as the usual activities, we also did the risk assessment we do for long-lining as part of the programme. Sue and I did it together with the first horse, talking through our process and explaining what we were doing to keep each other safe while we were assessing the horse. The second time Sue and I did it but with the students talking us through the process. Neither horse was safe enough for the students to long-line but that didn't matter as the process itself was useful. When we met up with Craig after the expedition, he said that the risk assessment was actually one of the most useful exercises as, when they got to Chile, much less had been put in place for them than was promised and so they had to do more for themselves – and make more judgements about risk and safety.

This workshop was of special interest to us as it was the first time we had worked with a participant who was profoundly deaf, using both hearing aids and lip reading. We quickly made the necessary and quite simple adjustments. It was an interesting exercise in energy. We noticed that when we needed to speak, we also needed to stop the exercise as the horse noticed the shift in energy when the young man had to look at us to lip read. We soon learned to give him a little bit more information at the start of each exercise so that he didn't have to stop and so lose the connection with the horse.

Apart from their help with logistics, the students looked after us very well. It was another opportunity to sample student rooms although this time in a much more traditional setting. I also managed to have a very quick look round Cambridge which has grown and changed hugely since my own student days there – although the traditional views of Kings College Chapel and the river were unchanged.

Back at NTU in the spring term, I was still reading a lot and wondering how to fit my project together. I shared my reading list with Carol but, to my disappointment, she never read any of it. She suggested that I should only be reading scientific papers but it was hard to find very much although several of the books had good bibliographies and these started to point me to more good sources. Most of the books were about energy in one way or another and that didn't seem to count as real science – neither did my demonstrations of matching horses to humans carried out with a pendulum! I had been explaining to Carol that the Limbic Stress Assessment system used the principles of kinesiology and the pendulum was a simple way of demonstrating this. Although I was loving the learning, it was becoming clear that there was no real understanding of what I was trying to do.

I had designed evaluation forms for our workshops some time before and was now very disciplined about having participants complete them. Even from the relatively small amount of data we had at the time – although backed up by a lot more informal feedback – I knew that the workshops had wide appeal. As I am an accredited personality profiling practitioner, I was interested in know what part personality type played in learning with horses. I was delighted when fellow practitioner Ali Stewart asked if a group from the South West of England could come up to Edinburgh and experience a day with the horses. What better group to give me detailed feedback about personality! We chose 1st April (All Fools Day in UK) as Ali and her group were able to get very good value airfares that day – but it also gave us lots of humorous ways of publicising the event.

We had a wonderful day in Edinburgh hosted by the Drum Riding for the Disabled Group at their very comfortable facilities. The group really enjoyed the learning and were brilliant at giving me detailed feedback especially as all of them knew their own personality types. It was also an opportunity for us to test out our latest activities which are based on the four main personality types. It was great fun watching the various combinations of horse and human personalities tackling each activity. Analysing the data afterwards did indeed show that our workshops provide valuable learning across all personality types. Although it was a small group, it showed that this was a line of enquiry worth pursuing. It was also quite fun to use my newly found statistical skills to do the analysis.

By now I was starting to prepare my end of year assignments – a one thousand word conference abstract, a fifteen minute conference presentation on the abstract topic and a project proposal which in my case would be a dry run for submitting my project approval form which I would need to do by the Autumn. I also had to do a short progress presentation on my project overall for the School's Post Graduate Conference.

In the spring, I had received the Call for Papers for the NARHA (North American Riding for the Handicapped Association) Conference to be held in November 2008 in Hartford, Connecticut. I decided to use this opportunity to submit an abstract for a real conference. I had already had an article about Measuring the Impact of Horses on Humans accepted for publication in the summer edition of the EFMHA journal and so I decided to have a go and submit an abstract along similar lines for the conference. (EFMHA is the Equine Facilitate Mental Health Association, a sub group of NARHA and, in spite of its name, has an interest in a wide range of learning and therapy with horses.)

As I was preparing my assignments, I was torn as to whether to use the conference abstract or prepare another one based on the work I had done on personality. I knew this latter topic would probably be better received as it was perceived to be more 'scientific' but on the other hand, I needed to know where I stood with my work and so I decided to use the real conference abstract and prepare a presentation which I could then lengthen if my submission for the NARHA conference was accepted. I prepared an abstract and presentation concentrating on Limbic Stress Assessment as by now I had a lot more information about how it worked and more about its 'scientific' pedigree.

The greatest challenge was the project proposal which I was doing along with a draft of my actual project approval form. I couldn't get my work to fit together other than as a rather disjointed list of studies. It was alright to a point but I didn't feel I was really doing my topic justice and, more importantly, it didn't make my heart sing either. I played around with the title to see whether that would give me a better focus – but it didn't make much difference. I was also struggling with my Literature Review. It wasn't that I hadn't been reading. On the contrary, I had read a great deal and probably become one of Amazon's best customers in the process. It seemed that this didn't count and I hadn't been reading the 'right' things.

In the end, I submitted what I thought was a rather lack-lustre project proposal for my assignment and an equally dull draft project approval form. To the best of my ability, however, they conformed to 'The Process'. I decided, however, that I would set NTU a test and do both my abstract presentation and my progress presentation from my heart. The response would determine my next steps.

Knowing that I would need my morning meditation more than ever, I had taken Linda Kohanov's 'Way of the Horse' cards with me to NTU to guide my thoughts for 'Presentation Day' – abstract presentation in the morning and progress in the afternoon.

As I contemplated my day, two cards came out together. The first card was Chiron. The gift of this card: "The organic synergy of instinct, intuition and reason gives rise to true brilliance". That alone gave me hope for my day – and there was more to come. In the chapter for this card, Linda Kohanov goes on to talk about the 'difficult' horse who can often teach us so much. "Sometimes by giving up our conventional ambitions and following this horse off the beaten path, we find we have, through much experimentation and soul-searching, learned something new we can teach to others." Of course I thought about Chelsea, the 'difficult' horse who had taught me so much, most of it well off the beaten track of conventional horsemanship. She would be with me as I made my presentations and I knew that whatever happened I would always have her company – and that being off the beaten track with her was usually great fun!

The second card was Boundary Dance. The message of this card: 'Holding your Ground.' As I read on, I began to laugh. "Kings who wield the power of life and death over their subjects have become managers who take advantage of employees, *tenured professors who put PhD candidates through hell (my italics),* doctors who shame and intimidate their interns and parents who treat their children like possessions." Linda describes this as a "demented form of authority."

I was ready for my day. I would hold my ground, knowing that my wonderful red horse was with me and that whatever happened there was still much for us to explore, to experiment with and to teach to others. It just might not be at NTU.

The presentation on LSA went quite well – although, despite my explanation, there were still a lot of blank faces from both staff and students. As I listened to the other presentations, I began to understand why. Most were about some physical attribute of the horse – paces, injuries, diet. They all followed the same pattern and they were all delivered in largely monotonous tones. I had always thought that people become involved in research because they are passionate about their subject but what I heard that morning made me wonder if research for some people was just a convenient means of delaying entry to the real world.

Although I had put heart and soul into my presentation, I couldn't help but remember the presentation I had done with Ann, Lynn and Su about LSA at the EGEA conference in January when we had used a wonderful slide show with beautiful photographs taken by Nancy Peregrine and Ann with background music and a commentary narrated by Su. It spoke volumes – and so we didn't have to say much other than answer questions.

By the afternoon session, I really did need the strength of the cards. It was possibly one of the most disappointing and disquieting conferences I have ever attended. I was third to go, following two other equine science students. Once more, their presentations left me feeling deeply saddened – not least for the horses involved in their research. One presentation concerned the measurement of fear/anxiety using an infra-red camera and the other was to investigate emotion in horses. Neither study took any account of the strength of the bond between the horses and the people involved in the studies. The clear expectation seemed to be that the horses would respond machine-like to whatever stimulus was presented, regardless of their connection to the humans assisting with the studies. It is entirely possible that there was no bond and that the horses had simply shut down, given the handling rules in force in the equestrian centre but I couldn't help but think about the very different responses the students might have received had they chosen to carry out their studies in a different environment. However, I couldn't help but notice that by excluding the bond, they made the science very easy.

I delivered my presentation to a sea of blank faces. My presentation contains many photographs of horses at liberty with people and I was aware that these caused an array of emotions – mainly negative and mostly disbelief that anyone could be so stupid as to let people and horses just *be* together. By the end I knew for sure that these 'tenured professors', lecturers and others in positions of power did not get what I was attempting to do.

I had also prepared a draft of my project approval form which I had sent to both Carol as my supervisor and to Jill Labadz as the school's co-ordinator of post-graduate studies. I had a long and uncomfortable meeting with Jill and a short uncomfortable meeting with Carol. The message from both of them was clear. Theirs was a science school and what I was doing was not science – not by their definition anyway.

I prepared to drive home. Although I had not finally made up my mind, I think I knew deep down that I was leaving for good.

Chapter 21: *Back into the Sunshine*

I returned home to the best possible news. My submission for the NARHA conference had been accepted and there was a copy of the EFMHA journal containing my article waiting for me. I reflected on how supportive the EFMHA research committee had been as I attempted to write my article in a scientific fashion. They asked many good questions, suggested where I might add in some extra information and generally helped me to polish up my paper. I learned more about scientific writing and experiment design from that experience than I did in my year at NTU.

There were workshops to be delivered for clients and also the first Equine Guided Coaches Circle (EGCC) to prepare for. EGCC was a step on from the research session we held in 2007. EGEA in January 2008 had been marred by atrocious weather with heavy rain and flooding which led to some of the sessions being cancelled. It was also very cold and so Ann, Lynn, Su, Aidan and I had opted to spend more time in our comfortable house sitting round the fire talking. Again we reflected that good as the conference was, we learned more from each other and thus the idea of EGCC was born.

We decided the group would be by invitation and we would extend these to people whom we felt shared our approach and values and who would be willing to share their learning. By September, we were a group of nine – Linda-Ann Bowling and Sandra Wallin from Canada,

Ann Romberg, Lynn Baskfield, Peggy Gilmer, Lissa Pohl and Susan Motzko from USA and Sue Hendry and I from Scotland. We didn't all know each other but we were all connected and it didn't take long before we were deep in conversation. Our two days together went in a flash. By the end everyone had contributed something and we had all enjoyed being coached by one another – something we realised was a rare treat for us as we are usually the coaches, at least with horses.

I was delighted to be able to collect more data from the LSA system and this time I gathered it in a more structured way. The learning from the horses was no less remarkable. When we did the first set of tests in 2007, we had used a round pen simply to make observation easier. In 2008, however, the round pen was no longer available and so we decided simply to spend the time with our allocated horse out in the fields, each with an individual observer. This made it clear beyond doubt that my experience with Gyp when he had sought me out was no accident. All our horses stayed with us until they decided their work was done. One participant was matched with two horses, Hooper and Moon. Hooper was near to the gate into the field and so she decided to approach him first. Moon was quite some distance away but immediately made his way down to join her and Hooper. It was such a deliberate move on his part. How did he know?

In 2008, I was allocated a pretty paint mare named Dove. When I approached her she was standing in the shade of a tree. During our session a couple of other horses arrived and looked intent on evicting us. I moved out of the way and Dove simply stepped round the other side of the intruders and met up with me again. It was an amazing little shimmy as I felt we maintained the connection throughout. After a little while, I became aware of Dove licking, chewing and sighing and realised that my session with her was at an end. Once again, the treatment had worked and my second LSA showed that everything other than my arthritis had come back into balance. It was interesting to see what the system had picked up this year. The only common feature was my arthritis but once more the diagnosis was accurate and reflected the hectic year I had had. The healing with Dove was so deep that I actually ended up with a cold – which was annoying as it slightly marred the few days I spent with friends in California before returning home. It was, however, interesting from a scientific point of view. For years now I have usually only had colds at the end of projects or other periods of intense work. It seems that as soon as I am able to relax, along comes the cold. After my treatment with Dove, I had that same feeling of really deep relaxation – and along came the cold!

I asked Su Wahl who was again administering the LSA process what she would recommend to help my arthritis and it was another opportunity to see the LSA system at work with a more 'conventional' treatment, in this case a neutraceutical. Sue recommended a product called Phenocane and when allocated to me on the system I could see the arthritis spike on the computer screen come back into balance. Then it was just a matter of fine-tuning the dosage and the system did that too. I've been taking it ever since and am currently enjoying the most pain-free winter ever although it has also been one of the coldest.

At the end of our Coaches Circle, we all voted it a success and decided to gather again in 2009, extend the membership to twelve people and the length to three days with each person leading a session. In October we published our proceedings so that we could all remember the learning.

While attending our Coaches Circle, I stayed with Ann Romberg and shared with her my experiences at NTU. It was good to have a neutral sounding board and wise listener. I was torn as to what to do. Leaving and pursuing my own studies until I found another academic home was an easy option but maybe part of my role was to go back to NTU and make a difference – if only for the sake of the horses? I didn't have to choose there and then and so I decided to sit and wait, watch and listen. Over the years another lesson I had learned from Chelsea was patience and living with not knowing, just seeing what showed up.

Once home, I started to prepare for the NARHA conference. As well as doing the presentation, I had also been asked to do a poster presentation – another new experience for me. The Coaches Circle had provided me with more data and as I prepared for the conference, the energy shifted and I found articles to read and ideas to pursue. I came to realise that much of what I was reading came from Quantum science and bioenergetics whereas NTU's view of science was purely traditional Newtonian science. Of course both have a place. Reading Amit Goswami's book 'The Quantum Doctor' helped me to understand how these two schools of thought can work together. I came to the conclusion, however, that whilst the Quantum scientists embrace Newtonian science, the reverse is not always true and this was what I was experiencing at NTU.

The more I read, the more I realised that my research journey was typical for someone at the edge of his/her field. I was hugely encouraged by books such as Candace Pert's 'Molecules of Emotion' and Bruce Lipton's 'Biology of Belief' – and also by the film 'What the Bleep do we know' in which Amit Goswami and Candace Pert appear, along with other new paradigm thinkers such as William Tiller, Joe Dispenza and Fred Alan Wolf. Candace Pert was a special inspiration. In her book she not only tells of her journey from being a very traditional scientist to what she herself describes as "my personal merging of East Coast science and alternative Californian 'healing' realms" but also of the difficulties and discrimination she experienced as a woman scientist.

More than anything, I was hugely encouraged by discovering work that had been done by scientists who wanted to apply their traditional rigour to new theories. As I mentioned earlier, I had always felt that the horses gave us much more than learning and now some of my clients were starting to notice this too. One primary school head teacher who came with a group of her pupils to spend a day with the horses told me "I so wanted to come with the children today because I knew it would be good for my soul." So what was it that the horses were doing? As I re-read Lynn McTaggart's book 'The Field', I came across the work of Elisabeth Targ. She had an impeccable scientific pedigree. Her father, Russell Targ, is a respected physicist who has received many accolades for inventions and contributions in lasers and laser communications. However, he was also involved in remote viewing experiments and was co-founder of the Stanford Research Institute's (SRI) investigation into psychic abilities in the 1970s and 1980s. Elisabeth was, as Lynn McTaggart puts it "a curious hybrid, drawn to the possibilities of her father's remote viewing at SRI but also shackled by the rigours of her scientific training." During the AIDS epidemic of the 1980's in California, Elisabeth was invited to design experiments to show the impact of remote healing on patients with such advanced AIDS that they were certain to die. The results surprised everyone – including Targ herself, who checked and rechecked the protocol and results several times to be sure they were correct.

Not only was it interesting to see how she designed and then carried out these experiments with scientific rigour but it also got me to thinking about whether the horses are actually delivering some form of remote healing during workshop sessions. Anecdotally, it seemed to make sense. As we had more repeat business, those returning for a second time began to be open about the 'feel good' factor they knew they would experience – as well as the high quality learning. Learning and wellbeing are closely linked with many studies showing that we learn best when we are feeling well and happy. Fear is the enemy of learning – which is why punishing school pupils for making mistakes is not helpful to their education. Perhaps people learned so well in our workshops because we allowed the horses to be healers and well as teachers. Suddenly my research was coming into focus and it also became clear to me that NTU was not the place to be and so I informed them that I would not be returning. The responses were polite although I think they were probably quite relieved to be rid of a student whom I am certain they perceived to be not really up to scratch.

Even with my extra knowledge, I was still a little apprehensive about how my presentation at the NARHA conference would be received. I worked hard to make it both interesting and informative whilst also being suitably scientific. In the end, I needn't have worried, the whole conference turned out to be highly enjoyable as well as offering some excellent learning. Best of all, those attending were wonderfully open-minded about everything that was presented whilst also asking some really good questions which helped everyone's learning.

I went out to Connecticut a few days early to take advantage of a visit to Green Chimneys Farm in upstate New York. Green Chimneys was founded sixty years ago and is an amazing school on a farm that offers incredible education and therapy to children suffering from mental illness, many of them from New York City. It was truly inspiring and the coach journey there and back was a good opportunity to get to know some of my fellow travellers.

I was fortunate to be speaking during the first session of the conference. My presentation was very well received and so I spent much of the remaining time having deeper conversations with people who were particularly interested in what I was doing as well as just enjoying talking to people with a shared passion for learning with horses. Halloween fell during the conference and I was delighted to be invited to a Halloween party at High Hopes Therapeutic Riding Centre by Kitty Stallsburg who is the Executive Director there. It was another wonderful visit to an excellent facility and we enjoyed a magical walk round part of the sensory trail, our way lit first by strings of fairy lights and then by torchlight. The conference finished with the Horse Expo held at the University of Connecticut's equestrian centre. I think this provided the highlight of the conference for me – watching a young girl who had been blind from early childhood ride her pony with total confidence and even tackle some jumps. Even more amazingly, we watched a video of her out trail riding with two sighted friends. Without being told of her condition, we would never have known she was blind.

I returned home feeling very at peace with where I was going with my research and looking forward to taking it forward more than ever.

Chapter 22: Findings to Date

Now that I was back from my travels, it was time to get ready for our own research workshops. I would be collaborating with Jan Young, a mature student who was in the final year of a sociology degree at Aberdeen University. I was introduced to Jan by a mutual friend who knew about my research and Jan's interest in non-verbal communication and horses. Jan explained that she was interested in our work with the horses and would like to use it as the subject of her final year dissertation. It didn't take us long to design a series of workshops that would give us each the data we needed for our respective studies. Jan was especially interested in body language and non-verbal communication and I wanted more on the learning outcomes and also on wellbeing. I was particularly interested to investigate three questions:

- Are there particular groups or personality types which benefit most from this form of learning with horses?
- Does initial anxiety about horses impact on learning outcome?
- Do horses have an impact on wellbeing?

Thanks to the amazing organisational skills of Lisa Esslemont, my wonderful PA, five workshops were arranged between 11th November and 11th

December, 2008, thus meeting Jan's very tight deadlines.

It was the first time that I was able to design a series of workshops purely for research purposes and it was fun seeing to what extent we could reduce the variables. All workshops were carried out in the barn at home with the same facilitators, the same horses, doing the same activities at the same time of day.

The workshops were free of charge as they were purely for research purpose and so it was a good opportunity to thank existing clients. Our invitations were taken up with much delight and only one company from my initial list declined. Interestingly, although I have worked with that company for some time, I haven't been able to persuade them to come for a session with the horses. Their reason for declining was that they felt it was inappropriate to be away from the office with a recession looming. The other companies on my initial list, all of whom had previously sent staff on workshops with the horses, appreciated some extra learning free of charge at this time – and several also mentioned the well-being effects they knew they would get which would be an added bonus with all the talk of a looming recession. The one refusal allowed us the opportunity to invite a potential new customer. Hopefully this will be good for business but, more importantly, it maintained the balance of participants that we wanted.

The participants, who were all volunteers, came from 5 organisations – a multinational oil services company, 2 professional services companies (one financial, one technical), an economic development company and a charity which sent four teenage clients and 2 members of staff. We explained to the participants before the workshops that these sessions were being run for research purposes and each group would be doing the same activities with broad learning objectives relating to leadership, team working and communication. This is the only significant difference from the usual workshops which are carefully tailored to meet client specifications.

There were 29 participants in total, 17 men and 12 women ranging in age from 16 to 49 years. Their education ranged from 5 who had school level qualifications to 6 who had post-graduate degrees. Most had no previous experience of learning with horses – or indeed any experience of horses at all. At the start of the workshop, 19 were completely comfortable with horses and 10 were mildly anxious. 19 of the participants knew their personality type and they were fairly evenly split across the four main types.

We decided to work with Chelsea and Susie who are both very accustomed to this work. We were quite curious to see what our horse partners would make of five identical workshops in a relatively short period of time.

Another difference from our usual workshops, were the number of professionals involved. I facilitated with Ruth Vaughan Hendry, a **red horse speaks** associate, as my horse person and Aidan took photographs. In addition, of course, we had Jan who was observing and we gave her Sue Hendry as her expert eyes. Our work was new to Jan and so having Sue interpret what was happening made the sessions much richer for Jan right from the outset.

Each workshop followed the same pattern:

- 9.30: Welcome, introductions and completing pre-session questionnaire

- 10.00 – 12.00 (approx): activities with the horses

- 12.00 (approx) till 1.00 (approx): completing post-session questionnaire, lunch and discussion.

The only exception to this was the group of young people from the charity who completed their questionnaires the day before and the day after at their premises to support their wider learning objectives.

Although the activities were the same for each workshop, we facilitated them at an appropriate level for the participants. With participants who are accustomed to personal development activities, we could use more of a coaching approach with open questions whereas for staff not accustomed to such events, our facilitation needed to be more direct. The facilitation also took account of the context of each organisation's work – for example, during the workshop with the group from the multinational oil services company there were discussions about communicating messages about safety and about safety leadership whereas with the professional services companies there was more discussion about communicating with clients and about personal effectiveness.

The results were fascinating. The figures for learning outcome were broadly similar to those collected from our routine workshops. The exception was for new learning which is perhaps not unexpected given that the research workshops were not tailored to client specifications.

However, the results suggest that, even if there was no new learning, the workshops were effective in embedding previous learning. This is borne out in comments from participants – for example:

"Reminded me of things I knew but didn't apply at work/home. This really sticks in my mind and will help use these skills in various settings"

"I enjoyed the session which has already helped some skill development".

We were also delighted to have confirmed that gender, educational attainment, job grade and personality type do not appear to have a significant impact on learning outcome. In other words our workshops provide good learning for everybody.

I had always wondered about the impact of anxiety/fear of horses on learning outcomes. I knew from conversations with clients that those who were anxious about horses at the start of a session were much more confident by the end but I was keen to investigate this further. Being able to do both pre and post session questionnaires was a perfect opportunity to include a question about this.

Again the results confirmed our informal findings. By the end of the workshop only 1 person was still slightly anxious about the horses (out of 10 at the start), although she added the comment "but more comfortable than before". Looking at the learning outcomes for these 10 participants showed that Initial anxiety about the horses does not affect learning outcome – or, indeed, enjoyment. All those who started being mildly anxious about horses actually scored 'A lot' in terms of enjoyment and the scores for learning effectiveness were similar to those who were fine about horses from the start. There is further work to be done to discover what contributes to the decrease in anxiety but from the work we have done since the research workshops, it is clear that it is the quality of the facilitation and our relationship with the horses that helps people to feel safe.

My greatest delight was that the results suggest that horses do indeed improve wellbeing. Everyone left feeling the same or better after the session with the horses. Of the 29 participants, 4 scored themselves on maximum 10 before and after the workshop and 1 scored 9 before and after. Of the remaining 24, only 2 scored no change. Of the 22 who scored an increase in wellbeing, 6 scored an increase of 4 or more points. There were no significant difference between males and females. Age, job grade, educational achievement and personality type were not significant.

We had deliberately not mentioned improved wellbeing as a possible benefit of attending the workshop although we did ask people to score themselves on this on the pre and post session questionnaires. Several people mentioned it by way of comment, on their post-session questionnaires.

"The combination of fresh air and working with the horses which was very peaceful" from a participant who went from a score of 9 to 10.

"Feeling better but still have things on my mind" from a participant who had been nursing a sick child and moved from a score of 2 to 7.

"A lot more relaxed than at the start" from a participant who went from a score of 3 to 8.

As well as the improvement in wellbeing scores noted in the questionnaires, there was also a marked change in body language by the end of the workshops in that it was softer and more expansive. It was also interesting to note that although the sessions were completed by 1.30pm to allow everyone to be back at work by 2.00pm, most people had to be 'reminded' to leave as they were so relaxed and settled!

The increase in wellbeing does not appear to be linked to overcoming anxiety about horses as the increases were spread across those who began feeling mildly anxious and those who were fine. The one person who remained mildly anxious about horses at the end of the session recorded a significant increase in wellbeing, moving from a score of 4 up to 8.

The other significant piece of feedback given by several participants was that the research we are doing is specific to **the red horse speaks** and shows why our programme is effective – and so we cannot yet say that learning with horses is always effective. It is no different from any other form of learning and can be delivered well or badly. The worrying fact is that when it is delivered badly it not only affects the physical and emotional safety of the participants but also of the horses. I hope in time we can develop some form of good practice guide based on our research. In the meantime, we will continue to deepen the learning we give to our **red horse academy** students so that they can deliver learning of the highest quality for the benefit of our clients and the wellbeing of both participants and horses.

Over the Christmas break following the research workshops, I continued to read and develop the next stages of my studies and am hopeful that I will find an interested and sympathetic academic home for them. As I looked back, I realised that I had done more in the four months since I left NTU than I had done in the year or so I'd been there. Once I was free of the need to squeeze my research into boxes which it didn't fit in the first place, the work just seemed to flow and, interestingly enough now has a better scientific base than ever before.

As I write this, I have just heard that I can continue my studies at Edinburgh Napier University in the School of Life Science. There is much excitement there about my work and there are great facilities for the work I want to do over the next few years to complete my PhD.

Part IV: *Magic*

Chapter 23: Magic Moments

Whilst it is important to follow sound business principles and establish a scientific base for this work, ultimately it is the magic moments that we and our clients remember. Here is a very small selection of our favourites. The clients' names have been changed for reasons of confidentiality.

"It's always the same. I can't keep staff"

One of our earliest magic moments occurred whilst working with a senior manager from an oil services company. He was working with Susie, the objective being to have her loose jump a small obstacle – a metaphor for getting over obstacles in the workplace. James carefully herded Susie round towards the jump but as she lined up to approach it, he moved in front of her. Being a polite horse, she stopped. This happened several times until Susie finally became frustrated and instead of turning to the obstacle, she jumped out of the area in which we were working into the other part of the big arena. This was no mean feat as we had created the small area with a barrier of show jumps about 1.30m high. Susie cantered off down to the far end then, realising she was on her own, came back, jumped in the small arena again and went to stand with a different group of people.

"It's always the same," exclaimed James. "I can't keep staff." There followed an interesting discussion about why that might be and he was quick to admit that his natural style was to over-manage individuals.

He wanted to try again and so this time we told him that he could stay by Susie's shoulder – but go no further ahead of her than that. It is fine to guide staff and indicate what is necessary – and then get out of the way and leave them to get on with it. He did so and as soon as she had a clear line to the obstacle, Susie jumped it with ease. Then, much to our delight, she continued round the arena and cleared it again all by herself! She knew exactly what she was being asked to do.

James went away and wrote the word 'Shoulder' at the top of each page of his diary for the following month (it takes a month to set a habit). Although he had always known that he tended to over-manage staff and had been told as much in feedback from colleagues, it wasn't till he saw the impact it had on Susie that he was able to change his style.

"I wonder if that's what my staff would like to do but are too polite"

Peter came along on a leadership programme. He was very confident and full of his own ability. When I invited him to do an activity with Chelsea, he went striding up to her – only for her to turn her butt towards him and then look round as if to say "I don't think so". She was clearly not about to kick or do anything nasty but it was very clear that she had no intention of co-operating wit him. Peter stopped in his tracks and looked at me.

"What shall I do now?" he asked.

"Well, what will you do?" I replied.

We began to talk and the conversation became a discussion about the difference between 'Ask' and 'Tell'. Peter acknowledged that he was a 'Tell' person. Clearly, however, that wasn't working with Chelsea so he agreed to 'Ask' her instead. He approached her again, now with softer body language and a more collaborative air and this time, she turned to face him. They completed the activity quickly and easily.

Afterwards Peter said to me "I wonder if that (Chelsea turning her butt) is what my staff would really like to do but are too polite." He went on to say that the incident with Chelsea had shaken him quite profoundly. He had rarely had any feedback and as is often the case in such circumstances, his opinion of himself well exceeded that of his staff. It had taken Chelsea to deliver that message to him. As well as adopting the 'Ask' approach more often he also resolved to seek more feedback from those around him.

A Community of Leaders

I have the privilege of working with some very high performing leaders. This group came together from around the world. They would be spending the next few days working on their individual leadership skills but would also have some project work to do together.

Their activity was to have the horses move together round the outside of the arena whilst they stayed linked to each other in a circular shape in the centre of the arena.

It took them some time to work out how to operate easily as a group whilst staying within the rules and then, quite suddenly, they started to move as one turning up the energy so that the horses trotted and then cantered. Just as easily, they were then able to turn it down, bringing the horses back to walk and halt.

It was beautiful to watch the flowing movement as the horses changed pace and direction guided by the silent group in the centre. Watching closely, I could see how the leadership was moving effortlessly from person to person, each one knowing exactly what needed to be done.

When we paused to reflect, one of the group said. "That was amazing. We were a community of leaders." It was a perfect description of their work together. Each person was prepared to be both leader and follower to achieve the task.

Confidence

I had a call from a young woman named Anna who had been referred to me for coaching. She was having a difficult time at work due to the behaviour of a colleague. She dreaded his phone calls and wanted to run away when she saw his number appear on her telephone.

Looking at her body language, I could see her lack of confidence. Although she had come along for conventional, conversational coaching,

I suggested to her that she might benefit from working with the horses. Despite her scepticism, she agreed to try this approach.

I introduced Anna to Chelsea and asked her to take Chelsea for a walk round the field. Chelsea very gently but firmly took charge and led Anna round the field. We stopped and I gradually helped Anna open her shoulders and raise her head, looking ahead to where she wanted to go. Immediately Chelsea sensed the change and allowed Anna to lead. A broad smile lit up Anna's face. "I'm in charge," she said, now leading Chelsea confidently wherever she wanted to around the field.

We talked about how Anna could take this confidence back to the workplace. The next time the dreaded number appeared on her phone Anna simply stood up and imagined Chelsea beside her. It worked a treat and the difficult behaviour disappeared. Anna came out for a couple more sessions to embed the behaviour and I was delighted to receive an email from her a few weeks later saying that she had been promoted.

I could have continued the conversational coaching and explained to Anna how to improve her posture and body language but ten minutes with Chelsea was so much more effective – and smile on Anna's face made it all so worthwhile. Experiential learning indeed!

"I can't get past the second item on the agenda"

When she held the weekly conference call with her sales team, Sarah could never get past the second item on the agenda. There was always some distraction or protracted discussion that meant the hour was up.

We chose long lining to help solve this problem as Sarah was at some distance from her team. We then set up five cones to denote the five items on the agenda, Sarah having told us that there were usually five items to discuss.

Sarah started off confidently past the first cone and on to the second one, at which point the horse wandered off to the side of the arena taking Sarah with her! We couldn't believe it – it was exactly what happened on the calls. Sarah struggled to get the horse's attention and take her on to the second cone. Once past this point Sarah completed the course quite easily. We requested that she complete the course again. This time she got round all five cones smoothly.

"What was the difference that time?" we asked her. "I kept my focus all of the time." In our discussion, Sarah realised that she found conference calls difficult and tended to lose focus after a few minutes, allowing her colleagues to take over. She acknowledged that she had done exactly the same with the horse on the first time round the cones. She realised that the calls were an effort for her and she needed to prepare well before them and stay in control throughout if they were to be useful to her.

You *will* smile

Jane came to us with a group who were on a programme for young people who were being supported to return to work or education. At the start of the day she would not even look at us or tell us her name. Chelsea adopted her, seemingly determined to make sure that Jane would enjoy her day. By the end of the day Jane had grown in confidence and decided she wanted to join in an exercise which involved leading the horse over and around poles on the ground.

We typically work with Susie for that exercise as Chelsea had often been quite nervous about poles on the ground and tended to rush. Jane was adamant, however, that she only wanted to do it Chelsea. We adjusted the poles a little, explaining to Jane that this was for Chelsea's benefit rather than hers. Chelsea walked over and through the poles as if she had never had a problem and the smile on Jane's face was of pure delight. The smile on Chelsea's face told us that she too had achieved her objective.

Hidden talents

During a visit by some primary school pupils, we demonstrated the different paces of the horses. The children had fun counting the number of hooves on the ground at walk, trot and canter. The highlight, however, was watching Susie loose jumping. They were quick to notice that as she landed, there was a moment when all her weight was on one hoof.

"How much does she weigh?" they asked.

"How much do you think?" I responded.

"Five hundred kilos" was the instant reply from a small voice at the back.

"How did you know that", I asked, surprised by the accuracy of the answer.

"Oh, I just knew," he said, shrugging his shoulders and blushing slightly.

This small boy was a farmer's son who loved nothing more than helping his dad. Even at the tender age of ten, he had seen many cattle go through the sale ring and already had a keen eye for an animal. Guessing Susie's weight so accurately was his moment of glory and an opportunity for praise that didn't often come his way at school, despite his teacher's best efforts.

Seeing the real person

Caroline came with a group of young people. She was full of fun despite having a disability that limited her mobility. Although she walked unaided, she was very stiff and at times almost seemed to lose her balance. She was determined to join in all the activities with her team mates and so we decided just to keep a close watch on her to be sure she stayed safe. When it came to leading Chelsea, I knew I need not be concerned as Chelsea can be trusted to take care of people with disabilities of all kinds.

Imagine my concern as I saw them move very quickly away from me. I was about to step in when they turned to come back towards me. Caroline was grinning from ear to ear and I then could see that every so often Chelsea was leaning her shoulder in towards Caroline to help he keep her balance.

That was profound learning for me. I saw a disabled young person. Chelsea saw a fun loving, enthusiastic teenager – the person within.

Catching a horse

"How do you catch a horse?" asked a teenage boy. We explained that the horse had to be a willing partner in this process. We offered him the opportunity to catch Darcy, our large bay gelding. We don't usually use haltering as an activity but we knew it would mean a lot to Steve. He set off up the field with Sue who would be on hand to help with the headcollar. Darcy came to meet them and lowered his head obligingly to allow Steve to put on the headcollar and then they walked down the field together. Steve didn't say a great deal but was clearly delighted by his success – something that had been lacking in his life. As he walked round the field gently leading the big horse, he seemed to grow before our very eyes. Something so simple had become utterly profound.

"I have a choice"

The little group of teenagers began to lose their patience because Chelsea was not co-operating with them. As she walked away, I reminded them that they needed her to be in their group to complete the activity. To help them calm down and decide what to do next, I suggested they take a few deep breaths. That alone was enough to bring Chelsea back in the group and very quickly they achieved their goal. In discussion afterwards, I asked what they had learned from the activity.

"I didn't know I could do that" replied one member of the group.

"Do what?" I enquired further.

"Well, next time there's a fight outside the pub, I'm going to take a deep breath and walk away."

As we carried on the conversation, it became clear that, for the first time, this young man realised that he could make a choice for himself and didn't always have to follow the crowd.

Part V: *In Conclusion*

Chapter 24: *Programmes for Today and Tomorrow*

Since starting **the red horse speaks** almost five years ago, we have become ever more creative in designing and delivering workshops. We still hold fast to our core principles of always having a horse person and a people person with groups of up to twelve people and an additional person for larger teams of up to about twenty people. It's rare for us to work with group larger than this. The number of horses is more variable but it is rarely more than the number of facilitators. For leadership development or when working with young people, the groups tend to be smaller – usually six to eight participants – so that we can work on a more individual basis with them.

Our care for the horses continues to be based on the principle of not using any equipment unless it is 'safe in unskilled hands™'. In fact over time, we are finding that we need fewer 'props' of any kind as we have devised more activities that allow the horses to speak for themselves. The horses still have their rest days so that they stay fresh and continue to enjoy their work. Judging by the welcome given to all our visitors, we think we can safely assume this is the case!

Our principle of tailoring each workshop to suit the client's desired outcome also continues unchanged. The main difference now is that we have many more activities to suit particular themes – for example, we can deliver workshops to allow clients the opportunity to learn about personality types and emotional intelligence.

When we started out doing this work, a day with the horses was often delivered in isolation but increasingly we are delivering multi-day programmes where the day with the horses is part of a wider programme of learning. Often we deliver the whole programme but we also enjoy working with other training providers as we have a wide knowledge of leadership and team effectiveness models that enable us to ensure that the session with the horses fits into the provider's overall package of learning. The research workshops have also inspired us to design half day workshops which we hope will make it easier for more groups to come and spend time with us.

We have also developed ever more sophisticated ways of measuring what we do so that we can demonstrate its value and work with our clients to fulfil their learning needs. All of this also supports my research and I am profoundly grateful to our clients for being so willing to support this aspect of our work. Another benefit of the research is all the background reading and studying that goes along with it. This has helped us create rich learning environments for our clients which, together with the ability to measure the effectiveness of our work, sets our programmes with the horses apart from others.

Although I love working at home with our own horses, I also enjoy working in other parts of the UK and overseas. We now have a good network of venues and horses which is allowing us to expand to meet client needs. We do not need specially trained horses and have our own assessment routine when we meet new horses – or those we have perhaps not seen for some time. While the horse person checks out the horses from a safety point of view, I enjoy devising the metaphors that we can use in the workshop to meet the client's learning outcomes. All horses have something to offer!

We have also continued to enjoy working with young people and have developed a strong working relationship with the staff from the Apex's Aberdeen Office. This charity works with young people who have offended, or are at risk of offending, to help them get back into work or education. We were especially delighted when some of the young people who had come out for a day with the rest of their programme group volunteered to come back again to help us with our research. We hope that by continuing to measure what we achieve, they and other similar organisations will be able to attract more funding so that more young people can benefit from learning with the horses.

Since we moved to our farm, we have continued to use Loanhead for most of our sessions as the large indoor arena means that we do not have to worry about the weather. However, we have recently brought clients out to the farm with very positive results. We still have an immense amount of work to do here to develop our learning facilities but we have been pleasantly surprised by how much people love being here, simple as it is at present.

We are blessed to live here in Midmar. As well as the beautiful landscape dominated by the summit of Bennachie, this is also very ancient land. There has been a church here since Pictish times and there are many standing stones in local fields as well as the very famous Recumbent Stone and associated stone circle in the grounds of the present church. The remains of the old church and the standing stones are popular with many of our visitors. I love to think of all the generations of inhabitants who have lived here and thank them for the wonderful energy they have left behind.

I have no doubt that this beautiful location adds to our work and explains why people love coming here. Even when we are not directly working with the horses, it's always a popular meeting place – and, of course, all our visitors know that there will always be an opportunity to spend time with the horses if they want. We hope that over the years we will be able to expand our facilities whilst keeping the ancient energy of the place.

Looking ahead, I am excited at all the possibilities there are for this work to grow and deepen as we learn more about what really happens when people and horses come together to learn.

Chapter 25: *My Journey*

For me personally, **the red horse speaks** has been an amazing journey. When I began to ride regularly back in 1996, I was starting to think about my work life winding down. I knew I never wanted to retire completely and I had a hazy vision of continuing to do some consultancy work whilst having more time for hobbies – including, of course, riding. Chelsea changed all that.

By the time I was ready to launch the **red horse speaks**, I knew I had found a passion that will keep me busy for the rest of my life. Over the years, I have had many discussions with clients and fellow professionals about being 'in flow.' Mihaly Csikszentmihalyi, who first described the concept, suggests that this involves "being completely involved in an activity for its own sake... Your whole being is involved, and you're using your skills to the utmost". I realised that most of the time I am working with the horses, I am in this state. It is definitely me at my best – for my clients and for the horses.

Undoubtedly the passion was ignited from the moment I met Chelsea but it started to burn brightly during 2000 when we attended the Secretan Gathering and, later that year, the Associate Leadership programme. I knew then that my future lay with learning with horses and set about making the necessary changes that would allow that to happen.

One of the major changes which I described earlier was the shift away from large change management projects to delivering more facilitation and coaching which could, of course involve the horses. In order to brush up my skills in these areas, I attended a couple of courses. At one of these, I was fortunate enough to meet Fiona Adamson, a psychotherapist who wanted to become a coach. We took an immediate liking to one another and decided to do some follow up coaching with each other which we still do today, nearly seven years later – half and hour each way every two weeks or so.

Fiona has been a huge support to me as I have grown and changed. From the start, she has helped me find a balance between doing and being; to stop rushing headlong from one task to the next (well – most of the time!). I have learned to become more contemplative, to have the confidence to bring the learning from Chelsea into my own daily life – and also to take time out to enjoy myself!

During my conversations with Fiona, I came to realise that I inhabit the edge, always seeing what is possible rather than accepting what is. On reflection, I realised that I have done so for many years. Looking back over my career, I have always been involved in innovation and change, working in Information Technology during the advent of desktop computers and the new ways of working they offered is but one example. Despite being accustomed to edge living, it was still a challenge and often felt a lonely and quite scary place to be and so I was glad to have the support of Fiona and the other likeminded souls who gradually came into my life.

One of the unexpected joys of this work has been the opportunity to travel. At first I was quite anxious for, although I had travelled and indeed lived abroad, I had always done so with Aidan. I soon discovered that being a solo traveller had pleasure all of its own and I began to look forward to my trips abroad. As well as making frequent trips to USA for conferences and the like, I was able to visit Australia for the first time in 2008 to help Graeme and Annie Phillips launch the first Leadership and Horses™ programme "down under". Yarrabin, their ranch near Bathhurst in New South Wales, is a magical place and a perfect setting for this work. I'm sure it will become a popular centre for leadership development with and without horses.

The move to our farm in 2005 was a very significant point in my personal journey as it represented a significant financial commitment and therefore a long term work commitment to fund it. Although it is a beautiful place to live and having the horses here is a great joy, there were times in the early days when it also felt quite daunting. Fiona introduced me to "Anam Cara" by John O'Donohue and his beautiful poem 'Beannacht' (Blessing) became a great source of inspiration and support, especially the opening stanza.

On the day when
the weight deadens
on your shoulders
and you stumble,
may the clay dance
to balance you.

It was a time of huge growth for me personally as I learned to stay centred and balanced even on difficult days and to feel secure in following my dream and indeed I did very often feel as if the clay danced to balance me. Over time, however, I have grown more comfortable as I become more certain that learning with horses is what I am meant to be doing. So many times when I feel anxious or afraid, some unexpected good thing happens – perhaps some new work comes my way or an interesting opportunity. It always feels as if the universe won't ever let me suffer too much! I have become better at noticing these things too and being grateful for these messages of hope and confirmation that I am on the right track. More than anything, I have learned to trust my intuition and follow my heart.

Recently when I was doing some learning with Elizabeth Harley of the Reiki Training Centre near inverurie, she said that the difference between change and transformation is that change is reversible, transformation is not. Her example was the caterpillar and the stages that it goes through to become a butterfly. It made me realise that I had certainly transformed as I could not imagine going back to the "ready to semi-retire" person I was in 1997. Although life is full of challenges, I love experiencing the magical connection that can exist between people and horses. I feel so very fortunate to have found my passion in life and to be able to follow it, guided and encouraged by my equine friends.

Many readers will know Ronald Duncan's Tribute to the Horse, a poem that is used at the close of the Horse of the Year Show held in England each year. I particularly love the opening words to this amazing tribute.

Where in the world can man find nobility without pride,
Friendship without envy or beauty without vanity?
Here where grace is combined with muscle
And strength by gentleness confined

For me it sums up the partnership that has existed between people and horses over the centuries, how our past has indeed been 'borne on his back' and how, throughout history, it has been 'his industry' that we have relied on.

My husband Aidan has added another verse to Ronald Duncan's poem which brings the relationship right up to date and expresses my hope for the future of people and horses.

And now, another gift
Transcending all of this.
Still she stands by our side
Exemplar of what can be -
A partner and a guide.
Hers, the truth of the moment,
Hers, the pulse of intuition,
Hers, the healing way.
She walks close beside us
Our future, now, with her shared
Through her, one with nature
And with her, better days.

Bibliography

All of these books and articles have influenced my thinking about learning with horses.

Dispenza, J. (2007) Evolve Your Brain. *Health Communications Inc, Deerfield Beach, Florida, USA.*

Edwards, P. Edwards, S., and Clampitt Douglas, L. (1998) Getting Business to Come to You. *Tarcher Penguin, New York, USA.*

Goleman, D., Boyatzis, R. and Mckee, A. (2002) The New Leaders. *Little, Brown, UK.*

Goswami, A. (2004) The Quantum Doctor *Hampton Roads Publishing Company, Charlottesville, Virginia, US*

Handy, C. (1995) The Empty Raincoat. *Arrow, London, UK*

Handy, C. (1999) The New Alchemists. *Hutchinson, London*

Kaplan, R. (2007) What to ask the person in the mirror. *Harvard Business Review, Special Edition, January 2007, pp86-95*

Jaworski, J. (1996) Synchronicity – the inner path of leadership. *Berrett-Kohler, San Francisco, USA.*

Kohanov, L. (2001) The Tao of Equus. *New World Library, Novato, California, USA.*

Kohanov, L. (2007) Way of the Horse. *New World Library, Novato, California, USA.*

Lipton, B. (2005) The Biology of Belief. *Mountains of Love Productions, Santa Rosa, California, USA.*

McCraty, R. (2003) The Energetic Heart: Bioelectromagnetic Interactions Within and Between People. *Institute of HeartMath, Boulder Creek, California, USA.*

McCraty, R., Atkinson M., Tomasino D., Trevor Bradley R. (2006) The Coherent Heart: Heart-Brain Interactions, Psychophysiological Coherence and the Emergence and System-wider Order *Institute of HeartMath, Boulder Creek, California, USA.*

McTaggart, L. (2001) The Field. Element, *London, UK*

Pert, C. (1997) The Molecules of Emotion. *Simon and Schuster, UK.*

Rashid, M. (2004) Life Lessons of a Ranch Horse. *David and Charles, UK.*

Rector, B. (2005) Adventures in Awareness. *Author House, Bloomington, Indiana, USA.*

Resnick, C. (2005) Naked Liberty. *Amigo Publications, USA.*

Roberts, M. Horse Sense for People. (2001) *Viking, new York, USA.*

Rock, D., Schwartz, J. The Neuroscience of Leadership. *Strategy and Business. Summer 2006*

Sams, J. and Carson, D. (1999) Medicine Cards. *St Martin's Press, New York, USA.*

Secretan, L. (1999) Inspirational Leadership. *Macmillan Canada, Toronto, Canada.*

Sheldrake, R. (1988) The Presence of the Past. Random *House Group, London, UK.*

A Message from the Author

I hope you have enjoyed this book and that it inspires you to follow your dream and make it a reality – whether it's with horses or is something completely different.

To learn more about **the red horse speaks** and our programmes, please visit www.theredhorsespeaks.com. You can also sign up there for our free monthly newsletter.

Alternatively, you can email me with comments, feedback and questions at Beth@theredhorsespeaks.com

With gentleness and joy,

Beth and the horses.

Lightning Source UK Ltd.
Milton Keynes UK
14 November 2009

146213UK00001B/36/P